THE BIBLE IN CONTEMPORARY CULTURE SERIES

1

Editors
Aaron Rosen, Wesley Theological Seminary, USA
Casey Strine, University of Sheffield, UK

THE BIBLE AND GLOBAL TOURISM

Edited by

James S. Bielo and Lieke Wijnia

LONDON • NEW YORK • OXFORD • NEW DELHI • SYDNEY

T&T CLARK
Bloomsbury Publishing Plc
50 Bedford Square, London, WC1B 3DP, UK
1385 Broadway, New York, NY 10018, USA
29 Earlsfort Terrace, Dublin 2, Ireland

BLOOMSBURY, T&T CLARK and the T&T Clark logo
are trademarks of Bloomsbury Publishing Plc

First published in Great Britain 2021
This paperback edition published 2022

Copyright © James S. Bielo, Lieke Wijnia and contributors 2021

James S. Bielo and Lieke Wijnia have asserted her right under the Copyright, Designs and Patents Act, 1988, to be identified as Editors of this work.

Cover design: Charlotte James

Cover image: Mont Saint Michel bay © Arthit Somsakul/Getty Images

All rights reserved. No part of this publication may be reproduced or transmitted in any form or by any means, electronic or mechanical, including photocopying, recording, or any information storage or retrieval system, without prior permission in writing from the publishers.

Bloomsbury Publishing Plc does not have any control over, or responsibility for, any third-party websites referred to or in this book. All internet addresses given in this book were correct at the time of going to press. The author and publisher regret any inconvenience caused if addresses have changed or sites have ceased to exist, but can accept no responsibility for any such changes.

A catalogue record for this book is available from the British Library.

A catalog record for this book is available from the Library of Congress.

ISBN:	HB:	978-0-5676-8139-3
	PB:	978-0-5676-9840-7
	ePDF:	978-0-5676-8140-9
	ePUB:	978-0-5676-8142-3

Typeset by: Trans.form.ed SAS

To find out more about our authors and books visit www.bloomsbury.com and sign up for our newsletters.

CONTENTS

List of Contributors   vii
List of Figures   ix

INTRODUCTION
   Lieke Wijnia and James Bielo   1

Part I
ORIGINS

Chapter 1
NAZARETH IN PEWTER:
PILGRIMS' BADGES OF LORETO, WALSINGHAM AND WAVRE
   Hanneke van Asperen   17

Chapter 2
'BLINDED BY THEIR ZEAL':
GUIDE BOOKS TO THE HOLY LAND
   Jack Kugelmass   38

Chapter 3
BACK TO THE GARDEN:
BRINGING VISITORS TO AMERICAN EDENS, 1885–1956
   Brook Wilensky-Lanford   62

Part II
PILGRIMAGE AND RELIGIOUS TOURISM

Chapter 4
THE LATTER-DAY SAINTS, THE BIBLE AND TOURISM
   Daniel H. Olsen and George A. Pierce   83

Chapter 5
LOOKING FOR A MIRACLE:
TOURISM, TANYA AND THEURGY AT THE GRAVE
OF THE 'LATE' LUBAVITCHER REBBE
    Simon Dein   104

Chapter 6
MEDIA PILGRIMAGE:
THE STORIES THAT SHAPE THE MODERN CAMINO DE SANTIAGO
    Suzanne van der Beek   123

Chapter 7
CULTURAL-RELIGIOUS ROUTES AND THEIR TOURISM VALORIZATION:
'IN THE FOOTSTEPS OF THE APOSTLE PAUL' IN GREECE
    Polyxeni Moira   147

Part III
HERITAGIZATION

Chapter 8
BIBLE MUSEUMS
    Crispin Paine   179

Chapter 9
REWRITING THE BIBLE:
THE VISUAL CULTURE OF CREATION SCIENCE
    Larissa Carneiro   201

Chapter 10
MUSIC, SCRIPTURE AND THE SACRED:
NEGOTIATING THE POSTSECULAR AT A DUTCH ARTS FESTIVAL
    Lieke Wijnia   221

Chapter 11
BUILDING ON THE GOSPEL:
THE MORAVIAN SETTLEMENT AT CHRISTIANSFELD
    Marie Vejrup Nielsen   242

AFTERWORD
    James S. Bielo   261

Index   271

Contributors

Hanneke van Asperen is Researcher at Radboud University, the Netherlands

Suzanne van der Beek is Assistant Professor at Tilburg University, the Netherlands

James S. Bielo is Associate Professor of Anthropology at Miami University in Oxford, Ohio, USA

Larissa Carneiro is Instructor of Religious Studies at Duke University, USA

Simon Dein is currently Visiting Professor and Senior Lecturer at Goldsmiths University of London, UK

Jack Kugelmas is Melton Legislative Professor and Director, Center for Jewish Studies, University of Florida, USA

Polyxeni Moira is Professor at the Department of Business Management /Tourism and Hospitality Management Sector at Piraeus University of Applied Sciences, Greece

Marie Vejrup Nielsen is Associate Professor at Aarhus University, Denmark

Daniel H. Olsen is Associate Professor in the Department of Geography at Brigham Young University, USA

Crispin Paine is an independent scholar, and Honorary Lecturer at University College London, UK

George Pierce is Assistant Professor of Ancient Scripture at Brigham Young University, USA

Lieke Wijnia is curator at Museum Catharijneconvent in Utrecht, the Netherlands, and a fellow with the Centre for Religion and Heritage at the University of Groningen, the Netherlands

Brook Wilensky-Landford is a PhD student in Religious Studies at the University of North Carolina in Chapel Hill, USA

Figures

Figure 1.1. Miniature of Luke, in a book of Hours, possibly Northern Netherlands, tempera on vellum, ca. 1495. The Hague, National Library of the Netherlands, Ms 135 G 19, f. 27v. Image in the public domain.

Figure 1.2. Badge of Loreto, found in Nieuwlande, pewter, cast, fourteenth or first half of fifteenth century, 51 × 43 mm. Langbroek, van Beuningen family collection, 1111. © Medieval Badges Foundation.

Figure 1.3. Badge of Walsingham, pewter, cast, fourteenth century, 77 × 48 mm. London, British Museum, Britain, Europe and Prehistory, 1989,0113.2. © The Trustees of the British Museum.

Figure 1.4. Badge of Wavre, found in Nieuwlande, pewter, second half of fifteenth century, ca. h. 25 mm. Langbroek, van Beuningen family collection, 107. © Medieval Badges Foundation.

Figure 4.1. A map showing the major Latter-Day Saint religious heritage sites and routes in the United States. Source: Think Spatial, Brigham Young University.

Figure 4.2. A photograph of the eleven-foot replica of a statue of Jesus Christ by Danish artist Bertel Thorvaldsen called the Christus. (Source: the authors.)

Figure 4.3. A photograph showing the 'Special Witnesses of Christ' exhibit on the bottom floor of the North Visitors' Center. (Source: the authors.)

Figure 7.1. The Footsteps of the Apostle Paul (Missionary journeys). Source: Geraki (2006). Translated from Greek to English by the author.

Figure 7.2. The monument commemorating the event of St. Paul's disembarking in Neapolis was erected in 2000 in the courtyard of the Holy Church of Agios Nikolaos in the modern city of Kavala. Photo by the author.

Figure 7.3. Stakeholders in the recognition and award of the certification of a religious-cultural route.

Figure 8.1. Behind the scenes at a Bible-collection museum. Photo by the author.
Figure 8.2. The Bible World, Bangalore, India. Photo by the author.
Figure 8.3. Building the set for 'The World of Jesus of Nazareth', Museum of the Bible, Washington, D.C. Photo by the author.
Figure 8.4. Adam in the Garden of Eden, Creation Museum, Kentucky. Photo by the author.
Figure 8.5. Bible History Exhibits in Pennsylvania County, Philadelphia. Photo by the author.
Figure 9.1. Noah and the Ark, Sistine Chapel, Michelangelo, 1508–12, fresco. Image courtesy of Wikipedia in public domain.
Figure 9.2. Wilhelmus Goeree, engraver (1690). Voor-Bereidselen Tot de Bybelsche Wysheid, p. 248, vol. 2. Collection of the author.

Introduction

Lieke Wijnia and James Bielo

In summer 2019, the Dutch newspaper *NRC Handelsblad* published a heritage special, which included an article titled, 'The Church as "Slow Tourism"' (Pama 2019: 7). Individuals involved in policy-making and project management for a sustainable future for religious heritage in the Netherlands shared their insights, underlining the urgency of 'the church' as a category needing renewed emphasis. One individual recounted how church buildings and congregations are poised to offer what museums can only strive for: an immersive experience in a themed setting replete with meaningful objects and historic atmosphere. The article went further, arguing that declining religious literacy in the Netherlands should not prevent people from visiting churches. In fact, it continued, a lack of knowledge may even make visiting churches easier. Visitors unburdened by too much information or experience can simply enjoy the architecture, aesthetics and performance. And there you have it: the church as exemplary site of slow tourism.

Global tourism is a defining feature in the constellation of contemporary socio-cultural dynamics. Sacred sites – from temples and cathedrals to shrines and other commemorative sites – see large numbers of visitors. For example, Mont Saint-Michel, which adorns the cover of this book, is France's third most popular tourist attraction after the Eiffel Tower and the Palace of Versailles. It is, effectively, the most popular religious heritage site in France. It was approved in 1979 for UNESCO status, applied to both the historical abbey and the natural surroundings of the bay. Other popular, seemingly non-religious tourist sites – like museums or natural parks – are sacralized in various ways through practice and promotion. In turn, religious traditions, knowledge and practices are increasingly subject to negotiation in the context of mass travel. Sacred sites are also often regarded as places where the twenty-first-century traveler can disrupt the fast-paced routine of everyday life, finding a moment of calm and perhaps reflection. This pattern of religious sites attracting diverse travelers develops alongside the fact that many of these religious sites are still, in fact, part of ongoing traditions: lived religion side-by-side with lived secularism. How do these two modes of travel and dwelling co-exist? How are the liturgical needs of sacred sites sustained and communicated alongside the emerging needs of religiously diverse travelers? Such questions are a prominent challenge in the future management of religious heritage and valuable for the analysis of religious tourism and pilgrimage.

The Bible and Global Tourism brings together eleven case studies that explore diverse relationships between modes of travel and biblically engaged forms of practice, memory and performance. In doing so, this volume illustrates the power of the social life of scriptures (Bielo 2009); that is, the creative work that people make scriptures do amid institutional contexts of power, ritual and place-making (cf. Wimbush 2015). Contributors explore this theme across time (from the Middle Ages to the early twenty-first century), space (with particular emphases on the United States, Western Europe and the Mediterranean) and tradition (Protestant, Catholic, Mormon, Jewish and secular). Throughout, the theme of *biblical tourism* is shown to be a productive starting point for understanding how identities, ideologies and histories are constructed and negotiated.

'Biblical tourism', as a unifying heuristic, encompasses a diverse set of phenomena. Examples range widely, from an account of three U.S. men between 1885 and 1956 who claimed to have discovered the location of the Garden of Eden in the Americas (Wilensky-Lanford) to contemporary travelers on the Camino de Santiago who are inspired more by mass media representations of sacred pilgrimage than historically Catholic devotional pathways (van der Beek). Contributors appeal to various terms, including 'pilgrimage' and 'religious tourism', to engage what might be more precisely, if more clumsily, called 'biblically framed travel'; that is, travel oriented around scriptural texts, stories, tropes and interpretations. Here, 'biblical tourism' is not used to index travel to Israel-Palestine, which has been richly explored by other scholars (e.g., Bowman 1991; Kelner 2012; Kaell 2014; Feldman 2016). In hopes of complementing this existing work, the biblical 'Holy Land' appears in this volume in more de-centered, idiosyncratic ways. Contributors explore how the biblical 'Holy Land' is remembered (Kugelmass), replicated (Van Asperen), imagined (Wilensky-Lanford), featured in the periphery (Moira), marginalized (Olson and Pierce) and assembled and curated for new contexts of display and consumption (Paine).

To introduce this collection, we briefly outline three issues that animate the eleven case studies to follow and their engagement with themes of identity, heritage and authority. First, we address how this collection emerges from and responds to enduring debates about the relationship between tourism, pilgrimage and other modes of travel. Second, we address how relating to the past is an integral dimension for many of the cases presented. And, third, we address how biblical tourism reflects and re-creates conditions of authorization.

Modes of Travel

The most widely discussed continuum in the scholarship on religion and travel is that between the pilgrim and the tourist, echoing binaries of religion–secular and sacred–profane (e.g., Badone and Roseman 2004). Variously imagined diagrams place the pious pilgrim as an ideal type on one end of the spectrum, who undertakes travel driven strictly by religious conviction and focused on spiritual ends. On the other end of the spectrum stands the tourist as an ideal type, driven by secular convictions of visiting popular destinations and fulfilling bucket list desires. Somewhere in between stands the religious tourist, who negotiates various (at times, competing) motivations (Smith 1992). Such diagrams are an attempt to delineate different types of travel by means of the traveler's motivations, yet they also show the complexities of definitively proclaiming people to be engaged in one or another kind of travel. They serve to outline the range of potential standpoints, and in doing so demonstrate how difficult it actually is to categorically differentiate among these intersecting identities.

This tendency toward multi-valent categories has long been part of the study of tourism and pilgrimage. MacCannell (1973) argued that 'the motive behind a pilgrimage is similar to that behind a tour: both are quests for authentic experiences' (593). Turner and Turner (1978) argued that a longing for the sacred through ritual activity was always present in travel: '[A] tourist is half a pilgrim, if the pilgrim is half a tourist. Even when people bury themselves in anonymous crowds on beaches, they are seeking an almost sacred, often symbolic mode of *communitas*, generally unavailable to them in the structured life of the office, the shop floor, or the mine' (3–4). Work on the history of pilgrimage has also observed that faith-based travel has always included elements of commerce and leisure (e.g., Karst 2018).

Neat and distinct categories to pinpoint, differentiate and explain travel as either tourism or pilgrimage have always been elusive, and are perhaps increasingly so amid the socio-cultural dynamics of late modernity. Transformed by processes like secularization, migration and digital technologies, these societal dynamics have been famously characterized as liquid (Bauman 2004). Flexible identity formation is one expression of life in Zygmunt Bauman's liquid modernity. As he observed, 'In the brave new world of fleeting chances and frail securities, the old-style stiff and non-negotiable identities simply won't do' (27).

Amidst these fleeting chances and frail securities, global tourism, including travel identified as either secular or religious, has increased dramatically. In turn, scholars have argued for devoting greater attention to tourism in the study of religion, and, vice versa, for an expanded focus on religion in the study of tourism (e.g., Stausberg 2011). Figured as 'intersecting journeys', Badone and Roseman (2004) argued for recognizing the pluralized realities of tourism, travel, globalization and religion. This way of framing the matter reinforces the entanglement, mutual influence and contested nature of processes and sites where tourism and pilgrimage in the late twentieth and early twenty-first century meet and merge. As they write: '[R]igid dichotomies between pilgrimage and tourism, or pilgrims and tourists, no longer seem tenable in the shifting world of postmodern travel. A key basis for distinguishing between pilgrimage and tourism involves assumptions about the beliefs and motivations of travelers who undertake journeys to religious shrines' (2).

Another key indicator of tourism and pilgrimage being conflated is the continuous renegotiation and expansion of pilgrimage as practice. The notion of pilgrimage in both popular and academic writing is increasingly used to describe travel that is deemed intentional, purposeful and transformative (Greenia 2018), or 'touristic travel in search of authenticity or self-renewal' (Badone and Roseman 2004: 2). Accounts of *Star Trek* conventions, travel to Elvis' Graceland Memphis home and Jim Morrison's Paris grave and Vietnam motorcycle rallies are all equally at home in the field of pilgrimage studies (Porter 2004; Margry 2008). The concept of pilgrimage has even hit the study of literature, with an analysis of Herman Meville's novel *Moby Dick* as 'a story of the soul's pilgrimage towards redemption and Ishmael's "damp, drizzly November" in his soul as a metaphor for the holy longing' (Gentile 2009: 403).

In this volume, we take this acknowledgement of increasing entanglement between the sacred and global mobility as an invitation to broaden the scope of biblically framed travel. A wider conceptual and empirical berth allows for explorations of ritual creativity and unexpected forms of *holy longing*, which people give expression to while on the road and away from wherever they call home. Contemporary cases ranging from the practices of a Danish Moravian town becoming a UNESCO world heritage site (Nielsen) to navigating a Dutch arts festival (Wijnia), and an abundant variety of displays at Bible museums across the globe (Paine), are joined by medieval travels to Houses of the Annunciation in Italy, England and Germany (van Asperen), nineteenth-century Jewish travel to the Holy Land (Kugelmass) and modern claims of locating the Garden of Eden in the Americas (Wilensky-Lanford). Throughout, these case studies consist of entanglements between culture, religion and social conventions, entanglements which make them all the more relevant for understanding how the Bible inspires activities of travel, and, vice versa, how travel inspires divergent uses and understandings of biblical scripture.

Negotiating Histories

The examples analyzed here closely examine how biblical tourism works as a frame for relating to multiple pasts: scriptural and social, real and imagined. In doing so, they engage a broader problem of how religion works to construct and negotiate relationships with time (cf. Bialecki 2009; Bielo 2017). Temporality – that is, models for the nature of time, its passing and our relationship to it – plays an organizing role in composing religious life. As Paden (1988) writes, 'each religious world has its own past. Each has its own history' (75).

Comparatively, the problem of temporality is actualized in widely differentiated ways. We might recall influential work in the history of religions, like Mircea Eliade's notion of the 'eternal return' (2005 [1954]). For Eliade, a core difference between 'traditional and modern' life was the former's tendency to be planted in 'cosmic rhythms', always seeking attachment to a sacred history through rituals of repetition. The theme of repetition has been used as mimesis as well as return, such as Andreas Bandak's study of sainthood in Syrian Orthodox Christianity (2015). Bandak argues that the lives of saints circulate narratively in Orthodox communities as moral and spiritual exemplars that should be imitated, creating connective tissue between past and future. One widely recognized actualization of temporality is also future-oriented, namely the prophetic and preparatory work of apocalyptic communities. For example, Susan Harding's historical ethnography of American Protestant fundamentalists argues that apocalypticism is 'a specific narrative mode of reading history; Christians for whom Bible prophecy is true do not inhabit the same historical landscape as nonbelievers' (2000: 232). Harding's sense that religion creates a distinct inhabited landscape is echoed by Paden's proposal that religion is a form of world-making. As distinct lifeworlds, he argues that religion's time–space axis is its most important feature: 'religions are communities of memory more than they are collections of dogma' (1988: 78).

For this volume's interest in the relationship between religion and travel, temporality is actualized in diverse claims of accessing the biblical past and using biblical frames to engage others' pasts (namely, sub-cultural and national). In several cases, biblically framed travel is centrally about accessing an idealized scriptural past. The organizing aspiration is defined by intimacy, a desire to more richly and more viscerally connect with an imagined biblical lifeworld. In pursuing this desire to experience the biblical past, a range of media are mobilized. In Hanneke van Asperen's account of medieval pilgrimage to Loreto, Walsingham and Wavre, pilgrims sought scriptural intimacy through the architectural form of the Holy House of Nazareth. Be it the miraculously relocated house in Loreto or miraculously revealed replicas, these physical structures promise visitors that in and through them their spiritual attachment to the Holy Family will be deepened. Crispin Paine's comparative analysis of Bible museums – ranging from Anglophone contexts to India, Germany, the Netherlands, France and Israel – also illustrates how accessing the scriptural past organizes biblically framed travel. Distinguishing between the Bible as a text to be read and interpreted and the Bible as an object to be engaged, Paine shifts the media focus from emplaced replication (as with Holy House architecture) to collections of manuscripts and archaeological material. Through amassing and displaying biblical texts and related antiquities, these museums immerse visitors into a biblically saturated tourist frame and re-assert a familiar promise: that an intimacy with scripture will be gained by dwelling in this immersion.

The biblical past is not, however, the only past these chapters demonstrate to be at stake amid biblical tourism. In several cases, biblically framed travel is used to negotiate particular sub-cultural histories. For example, Daniel Olsen and George Pierce compare multiple forms of faith-oriented travel among Latter-Day Saints (LDS). LDS communities have created numerous religious heritage sites in the Americas, which work to legitimize both their expanded scriptural canon and their place as a recognized expression of American religious pluralism (cf. Esplin 2018). Marie Vejrup Nielsen also addresses the entanglement of heritage, memory and lived religious identity in her analysis of Christiansfeld, Denmark. The focus shifts from legitimation to preservation, as the extant members of the Moravian Church exist simultaneously as embodiments of continued tradition, tourist hosts and ritual practitioners. In her analysis of the annual Dutch arts festival *Musica Sacra Maastricht*, Lieke Wijnia illuminates a different temporal dynamic, which is the question of religious heritage as performed in a postsecular tourist frame. Organizers of the arts festival self-consciously deploy biblical themes in ways designed to signify sacredness and celebrate dimensions of Christian history, all the while maintaining a Dutch expression of secular national tradition.

Relating to the past is an integral dimension to the eleven cases collected here, though the form of travel varies as does the past in question. As travelers, hosts, workers and audiences, the actors featured in these chapters construct relationships with the complex, conflicted and multi-directional field of time. In doing so, they embody the kind of broad social processes Munn (1992) termed 'temporalizations of past time' (112) and Hirsch and Stewart (2005) termed 'historicity', or 'the ongoing social production of accounts of pasts and futures' (262). From medieval pilgrims to contemporary Bible museums, this volume demonstrates that the work of temporality is central to the performance of biblically framed travel.

Conditions of Authorization

A third and final organizing theme for this volume is how biblically framed travel reflects and re-creates conditions of authorization. Following Asad (1993), we understand authorization as a socio-historical process in which institutional and individual actors construct and corrode structures of legitimation and their affective attachments (cf. Lincoln 1994). In the chapters to come, readers will encounter legitimacy being claimed, internalized and challenged through multiple media: guidebooks to place; material souvenirs; maps; artistic performances; physical journeys; and, the multi-modal choreography of ideologically saturated museums.

The expressions of authority shown to be at stake here range from charismatic spirituality to media representations, institutional legitimation, ideological gazes and territorial claims. Throughout, these chapters exemplify Morgan's (2014) observation that power does not reside in single objects, places, or people. Rather, power is constituted and emanated by networks, composed of diverse actors, institutions and media. He writes, 'the structure that enacts sacrality is not something static like a holy piece of matter, but is rather an assemblage...the sacred is not a single thing, but is better described as what happens over time, built up by need, opportunity, rivalry, and sheer chance' (92). Pilgrimage destinations are important, but their efficacy is ultimately enabled and enhanced by the complex of travel infrastructures, pathways and frameworks that keep people flows flowing and mobility mobile.

True to modernity's form, one expression of authorization entails travel and place-making's intersection with scientific legitimacy. We observe this primarily in chapters by Brook Wilensky-Lanford, Crispin Paine and Larissa Carneiro. In her analysis of Eden-seekers, Wilensky-Lanford illustrates how structures of science (from geography to botany) are mobilized to support and explain claims that the famed biblical site has been located. Paine and Carneiro both illustrate how the cultural freight of 'museum', as a trustworthy institution and bearer of scientific fact, is used to present 'the Bible' to multiple publics. While Paine explores Bible museums' appeals to scientific archaeology, Carneiro argues that creationist visual culture performs in a scientific register despite the thorough antagonism and epistemological divide between modern science and creationism.

Another expression of authorization observed here orients around the power of institutional bodies to grant legitimacy to places and to their claims about the past. Marie Vejrup Nielsen and Suzanne van der Beek both examine tourist locales designated as UNESCO World Heritage sites. They illustrate that the UNESCO designation infuses added value into place and experience, though often in ways that depart from strictly religious authorization. The multi-valent meaning of heritage that results shapes both how travelers imagine and anticipate travel, and how visitors and hosts interact. Polyxenia Moira takes up a related case of institutional authorization, the development of the European Cultural Routes system. Focusing on the example of Mediterranean pilgrimages that channel visitors 'in the footsteps of Paul', she demonstrates that travel's incorporation into this form of authorization is a way of choreographing the experience. Designation as a European Cultural Route enables (and encourages) travelers to trust the official itinerary for where to go, what to see and, ultimately, how to be there.

As demonstrated by Vejrup Nielsen, van der Beek, Moria and Van Asperen, to designate a type of travel as pilgrimage is itself a form of authorization. To go on pilgrimage is to endow particular places, traditions, rituals, histories and experiences with legitimacy. This is perhaps best demonstrated by Simon Dein's chapter, which examines the practice of Chabad pilgrimage to the burial site of the late Lubavitcher Rebbe in New York City. Here, pilgrimage works as memorialization, heightening the rebbe's charisma even in death. Pilgrims tell stories of miraculous healing associated with their travel to the burial site, further authorizing the rebbe's writings, a kind of expanded scriptural canon for Chabad adherents.

The eleven cases collected here offer multiple examples of how biblically framed travel reflects and re-creates conditions of authorization. In some cases, like Dein's, travel bolsters the authoritative force of a scriptural canon. In others, like van der Beek's, travel expands the sites of authorization beyond scriptural tradition to mass media. And, in turn, media coverage (be it in literature, documentary, or Hollywood film) endorses the perceived legitimacy of the pilgrimage sites. As a network geared toward legitimating power, touristic pathways, infrastructures and frameworks are used to make claims about scripture as well as the identities, ideologies and histories performed through scriptural frames. Ultimately, this volume demonstrates that biblical tourism in its diverse expressions is a generative phenomenon for exploring the workings of authorization as a socio-historical process.

Volume Overview

The Bible and Global Tourism is divided into three sections, 'Origins', 'Pilgrimage and Religious Tourism' and 'Heritagization'. The first section presents three cases of biblical tourism set in diverse historical periods. The second section includes four chapters of travel-driven practices and their scriptural implications both on and beyond pilgrimage-tourism. And the third section features four chapters that discuss various types of sites where travel, scriptural negotiations and cultural heritage intersect. All of the chapters adopt a comparative, historically attuned stance toward the category 'Bible', recognizing that there is no stable 'Bible' across time and that any 'Bible' is a contingent, social, historical and ideological claim to textual legitimacy and authority (Malley 2004).

'Origins' recognizes that the phenomenon of biblical tourism has deep historical roots and insists that our comparative understanding is enhanced when we think across time as well as place. Hanneke van Asperen analyzes the material culture item and pilgrimage souvenir of pewter badges as they circulated in the Middle Ages in Western Europe. She is particularly interested in badges from sites in Italy, England and Belgium associated with the 'Holy House' of Nazareth. Jack Kugelmass takes up a different form of pilgrimage material culture: Holy Land guidebooks that proliferated in Christian, Jewish and secular networks in the late nineteenth and early twentieth centuries. Kugelmass reflects on how different classifications of travelers were both presumed and produced by the guidebook genre. Brook Wilensky-Lanford shifts the focus from the widely recognized 'Holy Land' to three efforts between 1885 and 1956 to claim the geographic location of the Garden of Eden, respectively at the

North Pole, and in the U.S. states of Ohio and Florida. Wilensky-Lanford argues that these divergent and idiosyncratic projects are less legible when read through modern ideologies of biblical literalism and more legible when understood as Enlightenment projects committed to ideologies of progress.

The 'Pilgrimage and Religious Tourism' section begins with Daniel Olsen and George Pierce's comparative analysis of ritualized travel within The Church of Jesus Christ of Latter-Day Saints (LDS). From the Holy Land to 'New World' sites associated with the Book of Mormon and memorial sites of displacement and settlement, LDS religious tourism engages the particular problem of constructing distinctive Mormon claims to place and scriptural tradition.

As Olsen and Pierce grapple with the diffuseness of Mormon heritage production, for his exploration of the Chabad movement Simon Dein uses the lens of place-making. As such, he positions his empirical focus in New York City's Brooklyn borough in order to grasp practices within the context of charismatic Judaism. Dein asks how Chabad pilgrimage to the burial site of the Lubavitcher Rebbe Menachem Schneerson began and how its development informs our understanding of the Chabad movement and the social life of scriptures.

The following two chapters in this section discuss contemporary implications of traditional pilgrimage trails. Suzanne van der Beek explores the impact of media representations of walking the Camino de Santiago and their impact on the experiences of contemporary pilgrims. While the context of the Camino is undeniably rooted in Catholic traditions and ritual practices, contemporary pilgrimage develops its own frames of meaning making. Mass media such as books and films play a crucial role in contemporary conceptions of what it means to be an authentic pilgrim and how a pilgrimage is deemed successful. A popular film like *The Way* (2010) or a book like *Ich binn dan mal weg* (2006) not only shape expectations of pilgrims before they embark on their journey but equally impact their experiences along the way. Van der Beek explores how a historically rooted pilgrimage trail is reimagined in media, which in turn structures experience on the trail itself. In her case-study, Polyxeni Moira analyzes how a heritage instrument launched by the Council of Europe impacts the marketing and valorization of constellations of historical (including religious) sites. This instrument, 'Cultural Routes', is seen to encourage intercultural dialogue through cultivating shared responsibilities for heritage sites along the identified route. In addition to routes that cover various European regions, Cultural Routes are also organized on a national level. Moira particularly explores the implications of a route initiated in Greece, which invites visitors to follow in the footsteps of the Apostle Paul. She discusses the necessity of a multi-disciplinary and multi-institutional approach in successfully establishing and operating a religious cultural route.

The final section, 'Heritagization', takes as departure point the notion of heritage and the process of heritagization which is a recurring feature throughout the volume. Religious history, objects and practices are simultaneously preserved and transformed in cultural sites like museums and festivals. Such heritage sites are equally informative about contemporary producers, visitors and contexts as they are about the historical origins of the collections they present to public audiences. Crispin Paine explores the use of the category 'Bible Museum' to study the global manifestation of displays of collections presenting some aspect of biblical history. These attractions collect scriptural manuscripts and/or archaeological objects from the Levant and display them in narrated exhibits. Most of these sites are affiliated with Protestant denominations or organizations, and reproduce a pervasive Protestant textual ideology: as 'God's Word' the Bible has been divinely preserved, protected and promoted throughout time. Paine observes how the ideological approach of the museums' founders and operators impacts the type of collections formed, modes of preservation employed, and presentations displayed.

Next, Larissa Carneiro examines another type of museum, those dealing with young-earth creationism. Based on fieldwork at the Creation Museum in the U.S. state of Kentucky, Carneiro demonstrates how this museum employs visual and rhetorical strategies used in natural history museums to persuade visitors of their theological vision. Two dominant scientific paradigms, Darwinian evolution and geologic uniformitarianism, are countered with creationist claims. This, in turn, works in tandem with the museum's ideology that everything written in the book of Genesis is a record of literal history. The chapter focuses on the role of visual culture in the display of the relationship between science and religion, through which the museum visually invites visitors to re-imagine what may not be literally present in scripture.

To complement these examples of museum practice, Lieke Wijnia presents the cultural presence of scripture in a case study of an arts festival in the Netherlands. The annual *Musica Sacra* festival in the city of Maastricht embodies the notion of the postsecular, in which religious and secular heritage meet and potentially merge into new understandings of the sacred. Through the use of annual themes, the festival committee aims to connect biblical references to topical affairs, with the aim of including but also moving beyond traditional scriptural resonances. As such, the festival manifests itself as a postsecular site, continuously negotiating religious and secular meanings, and inviting visitors to do the same.

Whereas Wijnia discusses a localized site that has the potential of a national outreach, in the final chapter Marie Vejrup Nielsen focuses on a localized religious community, which has been provided with the potential of a global outreach: the Moravian Church in Christiansfeld, Denmark. Founded in 1773 and enshrined by UNESCO in 2015, contemporary life at Christiansfeld involves the co-mingling of continued religious life among practicing Moravians, practices of heritage management and the demands of being host to religious tourists.

As the volume's editors, we have aimed for a richly diverse collection of cases with respect to religious tradition, touristic practices and geopolitical location. Chapters address multiple expressions of Protestantism, Catholicism and Jewishness, as well as references to secularity. With respect to tourist practices, multiple perspectives are included, ranging from the more macro level of religious, governmental and non-state institutions to the micro level of individualized traveler experiences. Local mediators like guides and caretakers also represent this individually engaged level. It is a reminder that although policy on tourism and heritage may be constructed at the institutional level, the daily practice of tourism is performed by diverse actors intersecting in both choreographed and improvised ways.

Conclusion

To close, this volume opens onto three broad areas of inquiry, in hopes of offering useful ways forward. First, the geographically and culturally comparative framework demonstrates how scripture continues to influence contemporary touristic practices, despite drastically changing societal (including religious) dynamics around the globe. Perhaps it is fair to observe that not despite, but because of, the conjunction of processes that characterizes the postsecular – secularization, diversification and spiritualization (Wijnia 2018) – contemporary religious travel has become more complex, creative and pluralized. Declining institutional affiliation in Western Europe does not entail a decline in religious tourism. On the contrary, societal and cultural transformations seem to function as a motor behind increasingly diversifying and creative sites of practice, in which the sacred may be gaining new faces but is certainly not absent.

Second, the cases examined here reach across time and space, and one contribution of this volume will be to advance comparative questions about the intersectionality of scripture, travel, identity and power.

Relationships across time, between biblical tourism past and present, are poised for consideration. How do historically non-extant forms, historically continuing forms and relatively recent forms compare with respect to issues of scriptural legitimations, identity negotiation, materiality and authorizing particular patterns of mobility? Questions oriented around the long *durée* are especially resonant because the sweep of time can matter greatly in certain cases of religious place-making. From the Camino to Christiansfeld, biblical tourism can be defined by the accumulation of stories, symbols, artifacts and discourses in a particular place in time.

Finally, this volume highlights a dialectical quality of the social life of scriptures. The creators, caretakers, visitors and places explored here do not merely reflect pre-established ideologies of scripture as they interact with destinations. Rather, there is a dynamic interplay between tourism and scriptural engagement. In the chapters that follow, we see a constant exchange among textual ideologies and hermeneutics and touristic constellations – visitor practices, infrastructures and materialities (e.g., guidebooks, pewter badges, museum displays, festival brochures). In this way, we suggest a formative role for biblical tourism in the history of modern scripturalizing, outsized to any quantitative tally of individual travelers or destinations.

Bibliography

Asad, T. (1993), *Genealogies of Religion: Discipline and Reasons of Power in Christianity and Islam*, Baltimore, MD: Johns Hopkins University Press.

Badone, E. and S. R. Roseman, eds (2004), *Intersecting Journeys: The Anthropology of Pilgrimage and Tourism*, Urbana-Champagne: University of Illinois Press.

Bandak, A. (2015), 'Exemplary Series and Christian Typology: Modeling on Sainthood in Damascus', in A. Bandak and L. Hojer (eds), *The Power of Example: Anthropological Explorations in Persuasion, Evocation, and Imitation*, 47–63, Malden, MA: Wiley-Blackwell.

Bauman, Richard (2004), *A World of Others' Words: Cross-Cultural Perspectives on Intertextuality*, London: Wiley-Blackwell.

Bialecki, J. (2009), 'Disjuncture, Continental Philosophy's New "Political Paul", and the Question of Progressive Christianity in a Southern Californian Third Wave Church', *American Ethnologist*, 36 (1): 110–23.

Bielo, J. S. (2017), 'Replication as Religious Practice, Temporality as Religious Problem', *History and Anthropology*, 28 (2): 131–48.

Bielo, J. S., ed. (2009), *The Social Life of Scriptures: Cross-Cultural Perspectives on Biblicism*, New Brunswick: Rutgers University Press.

Bowman, G. (1991), 'Christian Ideology and the Image of a Holy Land: The Place of Jerusalem in the Various Christianities', in J. Eade and M. Sallnow (eds), *Contesting the Sacred: The Anthropology of Christian Pilgrimage*, 98–121, London: Routledge.

Eliade, M. (2005 [1954]), *The Myth of the Eternal Return: Cosmos and History*, translated by W. Task, Princeton, NJ: Princeton University Press.

Esplin, S. C. (2018), *Return to the City of Joseph: Modern Mormonism's Contest for the Soul of Nauvoo*, Urbana-Champagne: University of Illinois Press.
Feldman, J. (2016), *A Jewish Guide in the Holy Land: How Christian Pilgrims Made Me Israeli*, Bloomington, IN: Indiana University Press.
Gentile, J. S. (2009), 'The Pilgrim Soul: Herman Melville's Moby-Dick as Pilgrimage', *Text and Performance Quarterly*, 29 (4): 403–14.
Greenia, G. D. (2018), 'What is Pilgrimage?', *International Journal of Religious Tourism and Pilgrimage*, 6 (2): 7–15.
Harding, S. F. (2000), *The Book of Jerry Falwell: Fundamentalist Language and Politics*, Princeton, NJ: Princeton University Press.
Hirsch, E. and C. Stewart (2005), 'Introduction: Ethnographies of Historicity', *History and Anthropology*, 16 (3): 261–74.
Kaell, H. (2014), *Walking Where Jesus Walked: American Christians and Holy Land Pilgrimage*, New York: New York University Press.
Karst, L. (2018), *Pilgrimage as Sacramental Ecclesial Practice*, PhD diss., Emory University.
Kelner, S. (2012), *Tours that Bind: Diaspora, Pilgrimage, and Israeli Birthright Tourism*, New York: NYU Press.
Lincoln, B. (1994), *Authority: Construction and Corrosion*, Chicago: University of Chicago Press.
MacCannell, D. (1973), 'Staged Authenticity: Arrangements of Social Space in Tourist Settings', *American Journal of Sociology*, 79 (3): 589–603.
Malley, B. (2004), *How the Bible Works: An Anthropological Study of Evangelical Biblicism*, Walnut Creek, CA: AltaMira Press.
Margry, P. J. ed. (2008), *Shrines and Pilgrimage in the Modern World: New Itineraries into the Sacred*, Amsterdam: Amsterdam University Press.
Morgan, D. (2014), 'The Ecology of Images: Seeing and the Study of Religion', *Religion and Society: Advances in Research*, 5: 83–105.
Munn, N. (1992), 'The Cultural Anthropology of Time: A Critical Essay', *Annual Review of Anthropology*, 21: 93–123.
Paden, W. (1988), *Religious Worlds: The Comparative Study of Religion*, Boston: Beacon Press.
Pama, G. (2019), 'De Kerk als "Slow Tourism"', *NRC Handelsblad*, 8 August, C7.
Porter, J. E. (2004), 'Pilgrimage and the IDIC Ethic: Exploring *Star Trek* Convention Attendance as Pilgrimage', in E. Badone and S. R. Roseman (eds), *Intersection Journeys: The Anthropology of Pilgrimage and Tourism*, Chapter 9 [E-book], Urbana-Champagne: University of Illinois Press.
Smith, V. L. (1992), 'Introduction: The Quest in Guest', *Pilgrimage and Tourism. Annals of Tourism Research*, 19 (1): 1–17.
Stausberg, M. (2011), *Religion and Tourism: Crossroads, Destinations and Encounters*, London: Routledge.
Turner, Victor, and Edith Turner (1978), *Image and Pilgrimage in Christian Culture*, New York: Columbia University Press.
Wijnia, L. (2018), 'Beyond the Return of Religion: Arts and the Postsecular', *Research Perspectives in Religion and the Arts*, 2 (3).
Wimbush, V. ed. (2015), *Scripturalizing the Human: The Written as the Political*, London: Routledge.

Part I

Origins

Chapter 1

NAZARETH IN PEWTER:
PILGRIMS' BADGES OF LORETO, WALSINGHAM
AND WAVRE

Hanneke van Asperen

> ...and the guide said: 'Since I notice you're a devout sightseer, I don't think it right to keep anything from you: you shall see the Virgin's very greatest secrets'.
>
> Desiderius Erasmus in *A Pilgrimage for Religions Sake*, first published in 1526. (Erasmus 1997: 640–1)

Pilgrimage and tourism are hardly the same, but there are interesting areas of overlap, as suggested in this chapter's epigraph. Both pilgrimage and tourism can be motivated by curiosity and a desire to see unknown sites, and this was the case even in the Middle Ages, and certainly with regard to the Holy Land. Many wanted to visit the places named in scripture where Jesus lived, taught and healed. The gospel writers mention a few villages around Jerusalem that were of importance in the life of Jesus, for example Bethlehem (Mt. 2.1; Lk. 2.4) or Bethsaida (Mk 8.22). Nazareth is identified as the place where Jesus spent his childhood (Lk. 2.39; Mt. 2.23; Mk 1.9). The Gospel of Luke elaborates most on Nazareth and also places the Annunciation there (Lk. 1.26-38).

Given these scriptural references, Nazareth holds a symbolically potent position in Christianity, and hence in pilgrimage. Pilgrims who had traveled to Jerusalem were also eager to visit Nazareth. And with the ever-growing popularity of the Virgin during the later Middle Ages, the attraction of Nazareth increased. Fortunately for many, a person did not have to travel to the Holy Land to visit Nazareth. Important cult sites, specifically Loreto (Italy) and Walsingham (England), drew large numbers of pilgrims, offering an alternative to the long and difficult journey to the Holy Land. But the two sites were more than proxy Nazareths because the biblical dimension was only part of their identity. Both places offered the pilgrim a glimpse of a distant Nazareth, but each presented the visitor with a different experience that was captured and propagated on the pewter badges that pilgrims took away from these places. Here, I use medieval pewter badges of Loreto and Walsingham to reflect on the 'image' and identity of the cult site where they were sold. Based on the case of pilgrimage to Wavre (Belgium) I argue that Loreto badges were mediators that stimulated appropriation and re-invention of this biblical cult.

Pewter Badges

Since the development of mass pilgrimage in the eleventh century, pewter badges cast in molds and mass produced became almost commonplace, especially in the northern parts of Western Europe (Rasmussen and van Asperen 2019). The tin and lead alloy had a low melting point and was easy to process. Each mold could be used to produce dozens of badges before it finally wore out or broke. Pilgrims wore these badges on their hats, clothing and bags to visualize the sites that they had visited and to be recognized as pilgrims. Others could see where the wearer had been and may have been motivated to take up the script themselves, making these badges instruments of advertising. With the badge, the wearer took something tangible from the site and the medieval traveler believed that they took something of the cult object with them. The badge was charged with meaning because it represented the saint or reliquary via its image, and its sacred aspect could be enhanced by coming into contact with holy relics (Lee 2005: 375–6; Bruna 2006: 183; Jones 2007). By channeling divine powers, the object could be put to use at home, extending its social life beyond memento and promotion.

The archaeological record of pewter badges demonstrates how these images of pilgrimage and veneration travelled widely across Europe (Stopford 1994: 63–5; also Kunera), and also how the images changed when pewter badges from one site were copied in altered form to become suitable identity markers for a different site (e.g. Koldeweij 2007: 131; van Asperen 2019: 86–8). Badges from Riga, for example, refer back to souvenirs of far-away Rocamadour, an immensely popular center of Marian veneration in the twelfth and thirteenth centuries (Andersson 1989: 55; Bruna 2006: 264, 280). Many sites sold badges that were modelled on souvenirs from Aachen that had grown popular in the fourteenth and fifteenth centuries (van Beuningen et al. 2012: 191).

Because pewter badges were made of such lowly material, scholars sometimes tend to regard them with disdain or even ignore them. Just because badges were made of cheap material, which meant that most people could afford them, does not mean that mediocre artisans were involved. Esteemed artists sometimes designed badges, for example painter and engraver Albrecht Altdorfer (ca. 1480–1538) in Regensburg (Winzinger 1963: 29–30). Some surviving badges display high artistic quality, indicating that it was indeed superior craftspeople who produced them. They did not necessarily cast badges themselves, but carved molds for others to use for production. Despite their materials, pewter badges were not only bought by the poor. Men and women belonging to the ruling elite also sought them, for example the dukes of Burgundy and their close relatives (van der Velden 2001: 236; Koldeweij 2006: 55–6; Spencer 1998: 12–13).

The production of pewter badges was often guarded closely, and deregulated only on exceptional occasions, such as feast days when so many pilgrims visited that the authorities could not keep up with demand (Cohen 1976). In the jubilee year of 1466, 140,000 badges were sold in Einsiedeln in a short period of two weeks (Koldeweij 2006: 18). In Regensburg, a copy after an icon of the Virgin painted by St. Luke started to perform miracles in 1519. During the first year of the cult, 27,000 badges were produced there (Raff 1984: 48; Wittstock 1998: 85–6). These were not enough to satisfy the demands of the 50,000 pilgrims, and people were said to have cried when they were not able to get hold of a badge. In 1520, the Regensburg church sold 109,198 pewter badges and 9,763 silver ones. Reports on the numbers of badges were probably exaggerated to stress the cult's popularity, but they underline the importance of badges for pilgrims (van Asperen 2013: 215).

The popularity of pewter badges as cheap, but highly valued, keepsakes makes them important objects of scholarly research: they had a large audience that transcended geographic, social and cultural boundaries. From their routes of circulation to their iconography, these badges raise questions about how cheap, mass-produced items were used to shape identity and propagate Marian cults in the late Middle Ages.

Nazareth

Gabriel's announcement to Mary that she would become pregnant is considered the moment of conception and therefore the moment of incarnation. The angel uttered the words 'Ave Maria gratia plena', and Christ took on bodily form inside Mary's womb; the Word became Flesh (Jn 1.14). Consequently, the Annunciation is regarded as a milestone in salvation history. James of Voragine, for example, starts his *Golden Legend* (*Legenda Aurea*) with a classic division of time (de Voragine 1995: 1:3). The *Golden Legend*, describing the lives of saints according to the chronological order of the annual feast days, would become immensely popular in the late Middle Ages. The first period, or the time of deviation, had started with the Fall; the second period, or the time of renewal, with Moses. The Annunciation and conception of Jesus marked the beginning of the third period, or time of reconciliation.

Travel to the Holy Land to encounter the places of biblical significance, such as Nazareth, can hardly be overestimated. After Constantine's edict of 313, which granted religious freedom to Christians, pilgrims began to visit the holy sites in and around Jerusalem (Marchadour and Neuhaus 2007: 93–4). When pilgrimage became a mass phenomenon from the eleventh century onwards, many wanted to see the house where the Virgin and Christ had lived. After Muslim rule was established in 1187, pilgrimage to many Christian sites of worship was still permitted and possible (Pringle 2012: 1–8). Latin priests and deacons were installed at selected sites, at the Holy Sepulcher, of course, and also at Nazareth, with permission of the Islamic ruler Saladin (r. 1174–93). His son guaranteed Christian pilgrims safe passage in 1207.

The situation was not stable, however, and conditions changed frequently, meaning that pilgrimage to the Holy Land was difficult and unpredictable. Yet the number of pilgrims does not appear to have declined dramatically during periods of unrest. When Sultan Baybar took Nazareth in 1263 and destroyed the church, immediate steps were taken so that pilgrims could still visit the site (Pringle 2012: 9). The composer of the *Chronicon Tielense* mentions that Duke Arnold of Gelre undertook a pilgrimage to the Holy Land in 1450 (Kuijs et al. 1983: 173–4). The Bruges merchant and diplomat Anselm of Adornes is known to have visited the Holy Land in 1470–1 (Platelle 1982). Writing about the pilgrimage made between 1481 and 1485 by Ghent alderman Joos van Ghistele, Ambrosius Zeebout describes at length what the travel group encountered at the site of the Annunciation in Nazareth. He tells us that there was once a church there but it had been destroyed, and there was now a chapel where the choir used to be, 'with an entrance on the side where the wall of the church used to be before it fell down' (Zeebout 1998: 123). Although it remained a costly, time-consuming and risky undertaking, many pilgrims made the journey.

The pewter badges that became increasingly popular in Western Europe after the twelfth century were never as ubiquitous in the Holy Land. Here, pilgrims preferred biblical landscape items such as palm-branches, bits of earth, scrapes of stone or drops of oil from lamps at the holy site (Bruna 2006: 28–32; Boertjes 2014: 169–71). These practices extended to Nazareth, as a reliquary box with stones from different sites in Palestine including Nazareth illustrates (Rudy 2011: 109–10). There is evidence of pewter ampullae produced in Acre in the thirteenth century and pilgrims could probably purchase these small flagons on their way to holy sites (Kötzsche 1988; Syon 1999). These small flasks do not have iconography that connects them with any particular place of pilgrimage and they seem to have found a wide application suitable for different occasions, depending on where the pilgrim went. Badges have been tentatively attributed to Nazareth in the past, but this has been refuted (van Asperen 2013: 217–20). In short, there is no proof that pewter badges or any similar object with an image in low relief were mass produced in the Holy Land.

Loreto

Mentioned in a document of 1320 as nothing more than a 'ruralis ecclesia' Loreto developed into a renowned site of pilgrimage, probably shortly after 1360 (Grimaldi 2001: 9–11; Vélez 2019: 13). According to popular legend, angels lifted the Nazareth house where the Virgin had lived and moved it, first to Tersatto and, a few years later, to a location near Recanati. It finally ended up on the site where it would remain. The legend was antedated to 1294 when Jerusalem had fallen to the Muslim community and Christian churches throughout the Holy Land were destroyed, including those in Nazareth. This narrative of the arrival of the Holy House in 1294 nicely filled the chronological gap between the destruction of Nazareth's churches – still recapitulated by pilgrims in the fifteenth century (e.g. Zeebout 1998: 122–3) – and the origins of the cult in Loreto, providing the emerging site of pilgrimage with a solid groundwork. The narrative includes a comfortable margin of years for the house's temporary location at Tersatto.

While it offered pilgrims an alternative to the less accessible Holy Land, Loreto also benefited greatly from the growing popularity of Rome as a pilgrimage destination which was stimulated – if not initiated

– by the papacy. After the fall of Acre in 1291 there would be no more attempts to re-establish Christian rule in the Holy Land and therefore no possibility of gaining the indulgences that had been connected with a crusade. Instead Pope Boniface VIII granted a plenary indulgence to pilgrims who visited Rome in the jubilee year of 1300 (Dickson 1999). In its wake, there followed the offer of many more indulgences for visits to other sites on the Roman peninsula. In 1375 Pope Gregory XI offered pilgrims visiting Loreto a limited indulgence of ten years (Grimaldi 2001: 11; Borchardt 2017: 199–200). The pilgrimage attractions in Rome focused on events from the New Testament, especially Christ's Passion, and Roman churches owned numerous relics connected to the Passion that pilgrims could visit and venerate (de Blaauw 1997; Miedema 2001: 833–8). With these developments in mind, Loreto would have served well as an annex to the Roman pilgrimage in a way similar to how Nazareth complemented pilgrimage to Jerusalem.

On their arrival, pilgrims wanted to see the Holy House and enter the space once occupied by the Virgin and sanctified by her presence. Mysteries and miracles also focused on the physical presence of the house itself, and several miracle stories about Loreto focus on the integrity of the house that would remain intact in the face of all sorts of attempts to remove parts or change it. People were struck down with high fever when they tried to take pieces of stone from the house (Hutchison 1863: 37–40). After the inhabitants of Recanati had covered the house with a brick wall for protection, they found that the new wall had shifted away from the inner sanctuary on all sides, exposing enough room for a boy to walk in between (Hutchison 1863: 26). The inside of the house was characterized by its humble appearance, especially in comparison to its extravagant added exterior, but this underlined its quality as a relic from the Holy Land. Its simple appearance underscored its provenance, much like the quotidian landscape items of stones and dirt taken from the Holy Land (Nagel 2010: 215).

The Holy House had a wooden statue of the Virgin and Child, datable to the fourteenth century, on the altar inside (Grimaldi and Sordi 1995: 15–16). The statue was probably already in place in 1383 when Elisabetta Giles de Chierchi left three ducats in her will to pay for a vestment *'pro ymagine nostri domini Jesu Christi quem retinet in brachijs sacra majestas nostre domine virginis Marie de Laureto'*. The statue may have replaced an older painted image, possibly an icon, that is mentioned in a document of 1315 when Loreto was not yet the cult center that it would grow to be (Vélez 2019: 13). The panel was woven into the narrative of the cult nonetheless. According to a document of 1470, angels had placed this image of the Virgin inside the Holy House (Grimaldi 2001: 9–10). The icon connected the Holy House with St. Luke who, according to tradition, not only wrote the most elaborate of all gospels on the Annunciation and youth of Christ but was also assumed to have made portraits of the Virgin while she was living in Nazareth. Miniatures preceding the gospel of St. Luke usually depict the evangelist painting the Virgin (Fig. 1.1).

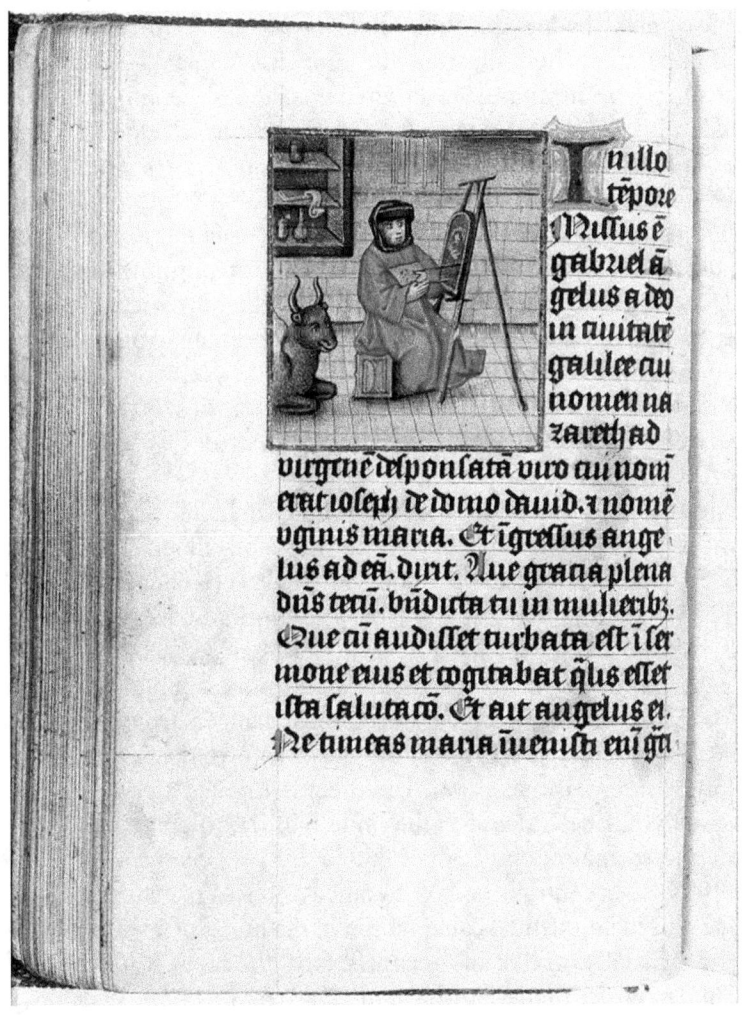

Figure 1.1. Miniature of Luke, in a book of Hours, possibly Northern Netherlands, tempera on vellum, ca. 1495. The Hague, National Library of the Netherlands, Ms 135 G 19, f. 27v. Image in the public domain.

With the growing number of visitors in Loreto at the end of the fourteenth century and into the fifteenth, badge production also increased. Late fifteenth-century administrative sources paint a picture of a lively trade in souvenirs, including badges, candles, votive offerings, rosaries, (woodcut) prints, possibly small parchment images and more. According to documentation from 1486, no fewer than forty workshops in Loreto paid their annual taxes (Grimaldi 2001: 471). Among these souvenirs were also metal badges in gold, silver or pewter. Depictions of Loreto badges in paintings help illustrate the popularity of small metal souvenirs at that time. On an altarpiece in Cremona, the Venetian painter Marco Marziale (1492/3–1507) depicted St. James with a badge of Loreto on his hat in 1507 (Pirovano 1985: 3; Penny 2004: 122–9). The Cremonese painter Lorenzo de Beci depicted St. Roch on an altarpiece in Binanuova in 1517, also with 'Loreto' sewn onto his hat (Kunera, no. 15471). Silver and gold badges must have been produced for pilgrims with more to spend, but these have not survived, as might be expected. However, pewter badges have been found over a large geographic area that extends beyond the Alps, to France (Arras and Paris), the Netherlands (Dordrecht and Nieuwlande) and the British Isles (Kunera, nos 00468–70, 01131, and 13789; British Museum collection online, reg. 1852,0427.1; PAS, NMS-ADABA2).

Some early Loreto badges survive that depict the Virgin and Child underneath a baldachin flanked by an angel on either side, perhaps in reference to the wooden statue in the Holy House (Kunera, nos 09039–40 and 12447–8). The iconography of the Virgin under a canopy with angels existed alongside that of the divine relocation of the house but was completely overshadowed by the latter in the course of the fifteenth century. Most surviving badges, and most badges from Loreto found outside Italy, depict the mystery of the divine relocation, focusing on the Holy House, an image that would come to characterize Loreto. The iconography has been dated to the middle of the fifteenth century, and it became increasingly popular over the course of a century (Grimaldi and Sordi 1995: 13). Central to it is the edifice, often in the shape of a church, with the Virgin and Child seated on top in the midst of its travel to Loreto (Fig. 1.2). Usually, one or more angels carry the house, lifting the Virgin and Child with it. Including the figures seems to imply that the Virgin herself came to the city of Loreto, suggesting that pilgrims who went to Loreto did not just visit a material building but were able to visit the Virgin and Jesus themselves.

Figure 1.2. Badge of Loreto, found in Nieuwlande, pewter, cast, fourteenth or first half of fifteenth century, 51 × 43 mm. Langbroek, van Beuningen family collection, 1111. © Medieval Badges Foundation.

Badges from Loreto have been found in Italy, France, the delta area of the Schelde and the British Isles. These finds can be augmented with a few Loreto badges that don't have specific find locations (Kunera, nos 04240 and 04407). Every badge, however, suggests the existence of many more because a single mold would have been used to produce hundreds before it wore out. The wide distribution of the few Loreto badges, both originals and depictions of them, indicates that Loreto attracted pilgrims from afar and that its image circulated widely.

Walsingham

Besides Loreto, pilgrims could turn to Walsingham to visit the Holy House. The legend about the origins of the Walsingham cult were written down in the *Pynson Ballad* that was printed in 1496, but probably written around thirty years earlier (Janes and Waller 2016: 4). According to the poem, the measurements and appearance of the Holy House were revealed to a noble woman called Richelde in a vision of the Virgin that had followed pious inquiry as to where she could best venerate the mother of Christ. Two locations qualified, each near a well, but after one was selected, workmen were unable to erect a building on the chosen spot. The next morning the material was found to have moved to the other location and this was believed to be an intervention of the Virgin, helped by angels.

The object of veneration (the house of the Annunciation) suggests a relationship with Loreto, and the date of origin (fifteenth century) indicates that the Walsingham cult was modeled after Loreto. The element of divine relocation from one site in Walsingham to another refers back to the cult in Loreto as well (Carroll 2016: 39), probably in an attempt to emulate the story of its model. The legend of Richelde was antedated to the thirteenth century to allow it to claim seniority over Loreto. Although the house was miraculously transported to a site indicated by the Virgin herself, it was a metric relic, an image modeled on a venerable object with the exact height and breadth of the original that it copied (Rudy 2011: 97–107). The measurements came from a trustworthy and verified source, the Virgin herself, and thus the house of Nazareth was established in Walsingham so that devotees could visit the site as if it were Nazareth. The shrine chapel was enclosed by a larger building, the Lady Chapel, separate from the priory, and its church was founded to oversee the shrine (Hearn and Willis 1996).

Figure 1.3. Badge of Walsingham, pewter, cast, fourteenth century, 77 × 48 mm. London, British Museum, Britain, Europe and Prehistory, 1989,0113.2. © The Trustees of the British Museum.

Badge production in Walsingham was not limited to badges depicting the Holy House (Fig. 1.3). Different types of pewter souvenirs featured images of the Annunciation or the Virgin and Child. Ampullae with the letter W (for Walsingham) or R (for Richelde) are also attributed to Walsingham (Spencer 1968: 140; 1971: 61–2; 1980: 10–19; 1990: 30–3; 1998: 135–48; Jones 2007; also consult PAS for many more examples). A badge depicting a monstrance with the text 'lac marie' [Mary's milk], and others resembling it, have also been attributed to Walsingham because the cult site was famous for its relic of the Virgin's milk (Spencer 1980: 14, no. 33; 1998: 147, no. 53a). The badges of Walsingham give insight into the various objects and images on display, and consequently into the composite nature of the pilgrim's experience in Walsingham. The construction at the cult site itself, where the Holy House was only part of the pilgrims' attraction, partially explains the variety of badges; there were different relics and saints to be venerated and there was a miracle-working statue of the Virgin that drew many pilgrims.

Carroll (2016: 36) has suggested that the badges of Walsingham might reveal a preference for certain relics and holy objects, but caution is in order. English badge finds depicting the Virgin, the Annunciation, or simply a lily are eagerly attributed to this popular site, even when the badge provides no direct evidence for such an identification (e.g. Spencer 1980: 12–13, nos 16–29; 1998: 143–5, nos 49–51; PAS, e.g. GFD_SF527947 and LON-D8EC47). The archaeologist and badge specialist Brian Spencer stated that the badges 'doubtless conveyed to those who wore them the particular association that the Annunciation was believed to have with Walsingham Priory' (Spencer 1990: 32). When Carroll (2016: 37) concludes that the images with the Holy House and the Annunciation were the most popular, he ignores the fact that many badges depicting the Annunciation (that is, without the Holy House) do not mention Walsingham and therefore might not have been produced there. Images like these were also worn as religious jewels, and may have been produced and sold to the wearers elsewhere by local craftsmen. Finally, the exact appearance of the miraculous statue that was burned in 1538 remains a mystery, but may also have been the subject of many badges (Kunera, nos 02155–7, 02424–34, 03156–61 and 11657). From other sources, the statue of the Virgin in the Holy House seems to have been at least as popular with pilgrims as the house itself. King Henry VIII inadvertently drew attention to the enormous popularity of the miracle-working statue when he ordered the cult image to be transported to London and burned, together with other images of the Virgin, as a public display (Dickinson 1956: 59–68).

The badges are illustrative of the range of objects on display, as mentioned in other sources, such as Desiderius Erasmus's 1512 colloquy called *Pilgrimage for Religion's Sake* (Erasmus 1997: 619–74). Here, Ogygius, who has been away on pilgrimage, describes his visit to Walsingham to Menedemus, who stayed at home. Despite Erasmus's satirical and ironical comments on the tourist industry that surrounded the cult in Walsingham (and other cults), the dialogue gives insight into a situation that the author had seen for himself. Not only does he mention the opulence – 'you would say it was the abode of the saints, so dazzling is it with jewels, gold, and silver' (629) – but also the objects on display for veneration: 'To the east is a small chapel, filled with marvels', among which are a joint of St. Peter's finger and the Virgin's milk (629). 'Thence on to the little chapel, the dwelling-place of the Holy Virgin [i.e. the Holy House]' (634), where Erasmus also places the miraculous image of the Virgin. He then mentions 'a small image [of the Virgin] displayed, unimpressive in size, material, and workmanship but of surpassing power...' with a jewel at her feet in the shape of a toad (639), 'gold and silver statues' and 'a world of wonderful things' on the altar (640). The badges illustrate the same range of venerable things, most prominently the statue and the various relics of the Virgin such as the house and her milk, in line with the indefatigable attempts to underline the Virgin's presence, both tangible and intangible.

Judging from the find sites of securely attributed badges, the popularity of Walsingham was limited to the British Isles. Admittedly, pilgrims from the Continent also visited Walsingham Priory, most demonstrably Erasmus, but they usually combined the journey with a visit to Canterbury, where St. Thomas Becket, who was considerably more popular on the Continent, was venerated (Koldeweij 2000: 49). Nevertheless, Walsingham was legendary in the British Isles. As Erasmus himself writes: '[The Virgin of Walsingham] has the greatest fame throughout England, and you would not readily find anyone in that island who hoped for prosperity unless he greeted her annually with a small gift, according to his means' (Erasmus 1997: 629–30).

Most pewter images of Walsingham's Holy House depict a scene of the Annunciation in its interior (Fig. 1.3). Gabriel and the Virgin are depicted under the roof of the edifice, on either side of the vase with the lily. Scrolls leave their mouths as the dialogue from the Bible is performed. The badges capture the moment that the Annunciation takes place, and it occurs inside the walls of the house in Walsingham. By means of the pewter badges, the Holy House in Walsingham was propagated as a relic of the Annunciation. Inside the chapel the devotees could contemplate the Annunciation, perhaps even experiencing the biblical moment of the incarnation as if they were present. A visit to Walsingham did not just equate to a pilgrimage to the Holy Land, but to a pilgrimage into biblical times, to the moment when the Annunciation happened.

Wavre

Finds of Loreto badges in France, the Low Countries and the British Isles bear witness to pilgrimages from northern Europe to Loreto (and back again). The Holy House's popularity with pilgrims from the Low Countries is also reflected in the souvenirs from a pilgrimage site of more than local renown, in a valley in the Duchy of Brabant called Basse-Wavre, or lower Wavre. According to legend, residents of the area heard angelic singing several nights in a row on special feast days of the Virgin, and also saw a divine light shining over the low-lying swamp beside their village. After the sick and destitute were miraculously cured, the inhabitants of Wavre planned a chapel on a hill nearby. Several times the work was resumed, but each morning the workmen found the bricks re-arranged in the valley below. One night, when they had decided to guard the construction site, priests saw the Virgin and some angels descending from heaven to carry the bricks downhill. The next morning, they also found a small, firmly closed box near the site where the Virgin had intended her chapel. The box was placed inside the chapel, because it was considered an *archeiropoeiton*, a relic not made by human hands: 'capsam non humano, sed angelico' (Wichmans 1632: 562). With its mysterious contents, it was also a shrine that functioned as a metaphor of the Virgin herself who had remained a virgin after conception. The box, both relic and reliquary, symbolized the Virgin's incorrupt body that had never been unlocked, its contents divine.

Featuring another divine relocation, the Wavre legend was closely modeled on the legend of the angelic transportation of the Holy House of Loreto. The chapel of Wavre was moved overnight from the top of a hill to a site in the valley indicated by the Virgin, just as the houses in Loreto and Walsingham were moved before they found their final destinations. The divine intervention to determine the site of the sanctuary in Wavre was coupled with the arrival of a divine reliquary that elaborates on the metaphors associated with the Holy House of Loreto and that also questions boundaries between relic and reliquary. The home of the Virgin, as it was propagated in Loreto, in Wavre became a symbol of her incorrupt body, the perfect shell that had encased Jesus. In Wavre, two important aspects of the Loreto cult – divine intervention and relic-reliquary status of the house – were pulled apart and became two separate strands of devotion. As a relocated Nazareth, Loreto was itself the subject of appropriation when the legend was borrowed and adapted to fit the new cult in Wavre.

Figure 1.4. Badge of Wavre, found in Nieuwlande, pewter, cast, second half of fifteenth century, ca. h. 25 mm. Langbroek, van Beuningen family collection, 107. © Medieval Badges Foundation.

Wavre badges were found in Haarlem (Fig. 1.4) and the flooded village of Valkenisse (Kunera, nos 00479 and 16652). Others without find location are kept in the royal library in Brussels (Hoc 1940). Although they are few, these badges give an interesting insight into the rhetoric used to identify this site. Their iconography corresponds closely with Loreto badges using the same visual elements. They take on a different shape: they are small pendants, in line with the trends of the late fifteenth and sixteenth centuries, as opposed to the larger open-work badges that were sold in Loreto in the fourteenth and fifteenth centuries. Badges from Wavre depict a reliquary-like box, sometimes carried by angels, sometimes with the Virgin seated on top of it (van Beuningen and Koldeweij 1993: 229; van Beuningen et al. 2012: 202). The miraculous event is pictured as a divine descent with the reliquary at the center. The elements were taken from Loreto badges, replacing the church-like Holy House (with bell-tower) with a box-shrine, and these were aligned almost identically using the same rhetoric of divine relocation. The close similarities suggest that the Loreto badges were used as visual models for the Wavre souvenirs.

The visual parallels between pewter souvenirs indicate that the Loreto badges may have played a vital role in mediation of the cult. Loreto badges were not copied without understanding the narrative that was coupled with it. The divine reliquary in Wavre was kept with several other relics that Godfrey I of Louvain (ca. 1060–1139), landgrave of Brabant, was supposed to have brought from the Holy Land. These included a fragment of the cross, a thread of the seamless robe, a hair of the Virgin, a pair of scissors and a needle that had belonged to the Virgin, as well as her waistband (Wichmans 1632: 566). Through these relics Wavre claimed a direct relationship with the Holy Land. These consciously superimposed connections with the Holy Land suggest a full awareness of the meaning of the Loreto cult and a conscious re-invention using the Loreto model to its full potential.

The legend associated with the origins of the Wavre cult was antedated to 1050, predating Loreto (Boone 1834: 9; Schoutens 1904: 131), as had been done in Walsingham, probably in an attempt to claim authority and veracity. It remains unsure when exactly the Wavre legend came into existence, but this was presumably in the fourteenth century. The archbishop of Malines mentions a miraculous statue in Wavre in a letter of indulgence dated 1698, where the image is said to have been miraculous for over two hundred years. Pewter badges that can be securely tied to Wavre because of an inscription are also dated to the end of the fifteenth century. The cult of the Holy House of Loreto had spread north across the Alps, probably primarily via the pewter badges that have been found in

France and the Low Countries and that predate the closely linked Wavre badges. The visual components of Loreto badges were imbued with new meaning, transforming the Holy House associated with the Annunciation into a divine relic of the Virgin.

Past to Present

Many medieval pilgrimage churches profiled themselves as houses of the Virgin. In theory, any church dedicated to the Virgin was a home of the Virgin (Harbison 1993), but Loreto and Walsingham were not just houses, they were relics. Their cults focused on a material building that had been central to the Virgin's life. At the same time, they were reliquaries. The walls enclosed a space where the Virgin had spent her time carry out her daily duties and prayer, but primarily where the angel Gabriel had visited her to announce her pregnancy and motherhood. Loreto and Walsingham were relocated biblical spaces for pilgrims to engage with outside of the Holy Land. Inversely, these relocated spaces made it possible for pilgrims to relocate themselves in a distant biblical past.

The location of the Holy House in Loreto has remained the same since the start of the cult. In Loreto, today, pilgrims seek the Holy House, although the biblical relic has been subjected to frequent transformation over the centuries, most notably with the addition of its marble encasing in the early sixteenth century (Vélez 2019: 52). The addition of the opulent marble cover underlines the double nature of the Holy House, which is both a relic – with a cover to protect it and to underline its spiritual value – and a reliquary covering the holy space inside. A space can only be experienced with boundaries, making the Loreto cult as much a cult of space as it is a cult of the shrine that encapsulates the room inside. The brick walls of the Holy House define the biblical space inside. At the same time, these walls are physical relics from the Holy Land much like the dirt and stones that people used to take with them from a Holy Land pilgrimage.

Much of the cult site in Walsingham was lost during the Reformation, although its remains have become part of a revival and its revived cult has received significant scholarly attention (e.g. Coleman and Elsner 1999; Coleman 2004, 2005; Waller 2011; Janes and Waller 2016). Despite the losses, Walsingham is still propagated as England's Nazareth. The ruins of the priory church can be visited today as part of a larger pilgrimage experience that incorporates surviving fragments of the old in a new, even larger, context involving a large part of the town, which has turned into 'a resonant three-dimensional frame within which experience appears to be amplified and given new significance' (Coleman 2004: 64). The cult space has widened from the roughly confined space of the Lady Chapel and well to a cult landscape incorporating much of the village and several different sites with different objects to visit. The miraculous statue, though destroyed during the Reformation, is present in the form of two replicas, one in the Anglican and the other in the Roman Catholic church (Coleman and Elsner 1999: 50). The Holy House was rebuilt using stones from dissolved and destroyed monasteries in England. Starting out as a cult of clearly defined space like Loreto, Walsingham has transformed into a cult of landscape, focusing as much on what was lost as on what is still there, or was re-invented.

Like tourists today, medieval pilgrims wanted to be spiritually enriched by the journey's experience and to take home a tangible memento. In the post-Reformation era, pewter badges were replaced by souvenirs in other media, such as prints. With the destruction of the cult images in Walsingham, souvenir production came to a halt (although it picked up again when the cult was given new impetus). Loreto has been the center of continuous attention since its beginning, and has continually produced souvenirs (Vélez 2019: e.g. 97, 124), although the material from which these souvenirs were manufactured changed with the invention of new methods and fabrication techniques. Iconography of the house's divine transportation by the hands of angels, as depicted on Loreto's pewter badges, found its way to other media and has remained much the same throughout the centuries (Grimaldi and Sordi 1995: esp. 20–6). This iconography was most likely established, distributed and popularized through pewter badges that, from the start, found a ready audience with pilgrims from far and wide.

Imitation and Emulation

The dynamics of pilgrimage profoundly influenced the religious landscape in Europe, giving rise elsewhere to new cults with new meanings in new contexts. Produced in large numbers and distributed over a large area, pewter badges are useful tools for studying the propagation and appropriation of cults. They show how cults were visualized, and they reveal nuances when it comes to the pilgrims' interests at these sites. The pewter badges from Loreto focus on the divine intervention that brought the house to Loreto, not on the Annunciation. Badges from Walsingham (which was also divinely relocated) focus instead on the house as the decor of the Annunciation. Thus, the journey to Walsingham was not only an alternative for a pilgrimage to the Holy Land, but a pilgrimage into biblical times.

The devotion to the Annunciation was multi-layered and the Holy House was venerated in different ways, through stone remnants of the biblical building (Loreto) and through copies, i.e. metric relics (Walsingham). Badges indicate how cults could be subject to re-invention, and which aspects were picked up for emulation. With the badges, the image of the Loreto cult traveled to northern Europe, where the biblical house became subject of repeated transformation. Both Walsingham and Wavre can be seen as appropriations of the successful Loreto cult, but each focused on different aspects. In Walsingham, the house was promoted as a relic of the Annunciation first and foremost, more so than had been the case in Loreto. The English cult was much more fragmented, however, and it resulted in a wide range of pewter souvenirs that visualize the variety within the Walsingham cult. Varied as they may seem, Walsingham badges concentrate on one connecting theme, namely the Virgin, and are instruments to underline her presence in Walsingham.

In Wavre the focus was not on the Annunciation, but instead centered around divine relocation and a divine reliquary. The underlying ideas of the Loreto cult were taken up in new ways: the connection with the Holy Land – not explicitly depicted on the badges – was resumed via additional relics, making the Wavre cult as much a site of biblical pilgrimage as Loreto. In Wavre, the relation with Loreto was visibly reinforced through the pewter souvenirs from the site, probably to benefit from Loreto's success and authority as Wavre tried to create a firm base for the newly established cult. Loreto was also used as an example to emulate: Walsingham and Wavre claimed seniority to Loreto. Although they modeled themselves on Loreto, both cult sites also deliberately distanced themselves from their immensely popular prototype, highlighting and re-inventing different aspects of (relocated) place, space and shrine. In Wavre, in particular, pewter badges were powerful mediators of these practices of relocation and appropriation.

Bibliography

Andersson, L. (1989), *Pilgrimsmärken och vallfahrt: Medeltida pilgrimskultur I Skandinavien*, Stockholm: Almqvist & Wiksell.

Boertjes, K. (2014), 'The Reconquered Jerusalem Represented Tradition and Renewal on Pilgrimage Ampullae from the Crusader Period', in M. Verhoeven, J. Goudeau and W. Weijers (eds), *The Imagined and Real Jerusalem in Art and Architecture*, 169–89, Leiden: Brill.

Boone, J. B. (1834), *Précis historique de Notre-Dame de Basse-Wavre*, Brussels: Vander Borght.

Borchardt, K. (2017), 'Late Medieval Indulgences for the Hospitallers and the Teutonic Order', in A Rehberg (ed.), *Ablasskampagnen des Spätmittelalters: Luthers Thesen von 1517 im Kontext*, 195–218, Berlin: de Gruyter.

Bruna, D. (2006), *Enseignes de plombs et autres menues chosettes du Moyen Âge*, Paris: Léopard d'Or.

Carroll, M. P. (2016), 'Pilgrimage at Walsingham on the Eve of the Reformation: Speculations on a "Splendid Diversity" only Dimly Perceived', in D. Janes and G. F. Waller (eds), *Walsingham in Literature and Culture from the Middle Ages to Modernity*, 35–48, London: Routledge.

Cohen, E. (1976), 'In haec signa: Pilgrim-badge Trade in Southern France', *Journal of Medieval History*, 2 (3): 193–214.

Coleman, S. (2004), 'Pilgrimage to "England's Nazareth": Landscapes of Myth and Memory at Walsingham', in E. Badone and S. R. Roseman (eds), *Intersecting Journeys: The Anthropology of Pilgrimage and Tourism*, 52–67, Urbana: University of Illinois.

Coleman, S. (2005), 'Putting It All Together Again: Pilgrimage, Healing, and Incarnation at Walsingham', in J. Dubisch and M. Winkelman (eds), *Pilgrimage & Healing*, 91–110, Tucson: University of Arizona.

Coleman, S. and J. Elsner (1999), 'Pilgrimage to Walsingham and the Re-invention of the Middle Ages', in J. Stopford (ed.), *Pilgrimage Explored*, 189–214, Woodbridge: Boydell & Brewer.

de Blaauw, S. (1997), 'Jerusalem in Rome and the Cult of the Cross', in R. L. Colella (ed.), *Pratum Romanum: Richard Krautheimer zum 100. Geburtstag*, 55–73, Wiesbaden: Reichert.

de Voragine, J. (1995), *The Golden Legend: Readings on the Saints*. Edited by W. G. Ryan, Princeton, NJ: Princeton University Press.

Dickinson, C. J. (1956), *The Shrine of Our Lady in Walsingham*, New York: Cambridge University Press.

Dickson, G. (1999), 'The Crowd at the Feet of Pope Boniface VIII: Pilgrimage, Crusade and the First Roman Jubilee (1300)', *Journal of Medieval History*, 25 (4): 279–307.

Erasmus, D. (1997), *The Colloquies of Erasmus*, translated by C. R. Thompson, Toronto: University of Toronto Press.

Grimaldi, F. (2001), *Pellegrini e pellegrinaggio a Loreto nei secoli XIV–XVIII*, Loreto: Tecnostampa.

Grimaldi, F. and K. Sordi (1995), *L'iconografia della Vergine di Loreto nell'Arte*, Exhibition Palazzo Apostolico della Santa Casa Loreto, Loreto: Cassa di Risparmio di Loreto.

Harbison, C. (1993), 'Miracles Happen: Image and Experience in Jan van Eyck's Madonna in a Church', in B. Cassidy (ed.), *Iconography at the Crossroads: Papers from the Colloquium Sponsored by the Index of Christian Art, Princeton University, 23–24 March 1990*, 157–70, Princeton, NJ: Index of Christian Art, Department of Art and Archaeology.

Hearn, M. F. and L. Willis (1996), 'The Iconography of the Lady Chapel of Salisbury Cathedral', in L. Keen and T. Cocke (eds), *Medieval Art and Architecture at Salisbury Cathedral*, 40–5, London: British Archaeological Association.

Hoc, M. (1940), 'Médailles de N.-D. de Basse-Wavre', *Revue Belge de Numismatique et de Sigillographie*, 13–19.

Hutchison, W. A. (1863), *Loreto and Nazareth: Two Lectures containing the Results of Personal Investigation of the Two Sanctuaries*, London: Dillon.

Janes, D. and G. F. Waller (2016), 'Walsingham: Landscape, Sexuality, and Cultural Memory', in D. Janes and G. F. Waller (eds), *Walsingham in Literature and Culture from the Middle Ages to Modernity*, 1–20, London: Routledge.
Jones, P. M. (2007), 'Amulets: Prescriptions and Surviving Objects from Late Medieval England', in S. Blick (ed.), *Beyond Pilgrim Souvenirs and Secular Badges: Essays in Honour of Brian Spencer*, 92–107, Oxford: Oxbow Books.
Koldeweij, A. M. (2000), 'Te Sente Thomas van Cantelberghe, in Inghelant...': Pelgrimsinsignes en pelgrimstochten naar Thomas Becket', in R. Bauer (ed.), *Thomas Becket in Vlaanderen: Waarheid of legende?*, 49–71, Kortrijk: Groeninghe Museum.
Koldeweij, A. M. (2006), *Geloof & Geluk: Sieraad en devotie in middeleeuws Vlaanderen*, Arnhem: Terra.
Koldeweij, A. M. (2007), 'Kreeg de onthoofde heilige ook een reliekhoofd? Over bedevaarten naar Sint Lieven, pelgrimstekens en reliekbustes...', *Handelingen der Maatschappij voor Geschiedenis en Oudheidkunde te Gent*, ns 61: 123–48.
Kötzsche, L. (1988), 'Zwei Jerusalemer Pilgerampullen aus der Kreuzfahrerzeit', *Zeitschrift für Kunstgeschichte*, 51 (1): 13–32.
Kuijs, J., L. de Leeuw, V. Paquay and R. van Schaïk (1983), *De Tielse kroniek: Een geschiedenis van de Lage Landen van de volksverhuizingen tot het midden van de vijftiende eeuw, met een vervolg over de jaren 1552–1566*, Amsterdam: Verloren.
Kunera. Nijmegen: Radboud University Nijmegen, https://www.kunera.nl/ (accessed 20 January 2020).
Lee, J. (2005), 'Beyond the Locus Sanctus: The Independent Iconography of Pilgrims' Souvenirs', *Visual Resources*, 21 (4): 363–81.
Marchadour, A. and D. Neuhaus (2007), *The Land, the Bible, and History: Toward the Land That I Will Show You*, New York: Fordham University Press.
Miedema, N. R. (2001), *Die römischen Kirchen im Spätmittelalter nach den 'Indulgentiae ecclesiarum urbis Romae'*, Tübingen: Niemeyer.
Nagel, A. (2010), 'The Afterlife of the Reliquary', in M. Bagnol, H. A. Klein, C. Griffith Mann and J. Robinson (eds), *Treasures of Heaven: Saints, Relics, and Devotion in Medieval Europe*, 211–22, Exhibition Baltimore, the Walters Art Museum, New Haven: Yale University Press.
PAS, Portable Antiquities Scheme. British Museum and Amgueddfa Cymru – National Museum Wales, https://finds.org.uk/ (accessed 20 January 2020).
Penny, N. (2004), *The Sixteenth-Century Italian Paintings, I: Paintings from Bergamo, Brescia and Cremona*, London: National Gallery Company.
Pirovano, C. (1985), *I Campi: Cultura artistica cremonese del Cinquecento*, Milan: Electa.
Platelle, H. (1982), 'Le pèlerinage en Terre Sainte d'Anselme Adorno (1470–1471) d'après un ouvrage recent', *Mélanges de Science Religieuse*, 39 (1): 19–28.
Pringle, D. (2012), *Pilgrimage to Jerusalem and the Holy Land, 1187–1291*, London: Routledge.
Raff, T., ed. (1984), *Wallfahrt kennt keine Grenzen: Ausstellung im Bayerischen National- museum, München 28. Juni bis 7. Oktober 1984*, Munich: Bayerisches Nationalmuseum and Adalbert Stifter Verein.
Rasmussen, A. M. and H. van Asperen (2019), 'Introduction: Medieval Badges', *The Mediaeval Journal*, 8 (1): 1–10.
Rudy, K. M. (2011), *Virtual Pilgrimages in the Convent: Imagining Jerusalem in the Late Middle Ages*, Turnhout: Brepols.

Schoutens, S. (1904), *Maria's Brabant, of beschrijving van de wonderbeelden en merkweerdige bedevaartplaatsen van Onze-Lieve-Vrouw in Brabant*, Aalst: De Seyn-Verhougstraete.
Spencer, B. W. (1968), 'Pilgrim Badges', in J. G. N. Renaud (ed.), *Rotterdam Papers. A contribution to medieval archaeology*, 137–53, Rotterdam: Coördinatie Commissie van Advies inzake Archeologisch Onderzoek binnen het ressort Rotterdam.
Spencer, B. W. (1971), 'A Scallop-shell Ampulla from Caistor and Comparable Pilgrim Souvenirs', *Lincolnshire History and Archaeology*, 6: 59–66.
Spencer, B. W. (1980), *Medieval Pilgrim Badges from Norfolk*, Hunstanton: Witley.
Spencer, B. W. (1990), *Salisbury Museum: Medieval Catalogue, Part 2: Pilgrim Souvenirs and Secular Badges*, Salisbury and South Wiltshire Museum.
Spencer, B. W. (1998), *Pilgrim Souvenirs and Secular Badges (Medieval Finds from Excavations in London)*, London: The Stationery Office.
Stopford, J. (1994), 'Some Approaches to the Archaeology of Christian Pilgrimage', *World Archaeology*, 26 (1): 57–72.
Syon, D. (1999), 'Souvenirs from the Holy Land: A Crusader Workshop of Lead Ampullae from Acre', in S. Rosenberg (ed.), *Knights of the Holy Land: The Crusader Kingdom of Jerusalem*, 111–15, Jerusalem: The Israel Museum.
van Asperen, H. (2013), 'Annunciation and Dedication on Aachen Pilgrim Badges: Notes on the Early Badge Production in Aachen and Some New Attributions', *Peregrinations*, 4 (2): 215–35.
van Asperen, H. (2019), 'Secular Power, Divine Presence: The Badges of Our Lady of Aarschot', *The Mediaeval Journal*, 8 (1): 79–107.
van Beuningen, H. J. E. and A. M. Koldeweij (1993), *Heilig en Profaan: 1000 Laatmiddeleeuwse insignes uit de collectie H. J. E. van Beuningen*, Cothen: Stichting Middeleeuwse Religieuze en Profane Insignes.
van Beuningen, H. J. E., et al. (2012), *Heilig en Profaan 3: 1300 insignes uit openbare en particuliere collecties*, Langbroek: Stichting Middeleeuwse Religieuze en Profane Insignes.
van der Velden, H. (2001), 'Karel de Stoute op bedevaart: De aanschaf van pelgrimstekens door de graaf van Charolais', in H. J. E. van Beuningen, A. M. Koldeweij and D. Kicken (eds), *Heilig en Profaan 2: 1200 Laatmiddeleeuwse insignes uit openbare en particuliere collecties*, 234–41, Cothen: Stichting Middeleeuwse Religieuze en Profane Insignes.
Vélez, K. (2019), *The Miraculous Flying House of Loreto: Spreading Catholicism in the Early Modern World*, Princeton: Princeton University Press.
Waller, G. F. (2011), *Walsingham and the English Imagination*, Farnham: Ashgate.
Wichmans, A. (1632), *Brabantia Mariana Tripartita*, Antwerp: Cnobbaert.
Winzinger, F. (1963), *Albrecht Altdorfer: Graphik. Holzschnitte, Kupferstiche, Radierungen*, Munich: Piper & Co.
Wittstock, J. (1998), 'Der Bremer Pilgerzeichen-Fund', in K. Herbers and R. Plötz (eds), *Der Jakobuskult in 'Kunst' und 'Literatur': Zeugnisse in Bild, Monument, Schrift und Ton*, 85–108, Tübingen: Narr.
Zeebout, A. (1998), *Tvoyage van Mher Joos van Ghistele*. Edited by R. J. G. A. A. Gaspar, Hilversum: Verloren.

Chapter 2

'Blinded by Their Zeal':
Guide Books to the Holy Land

Jack Kugelmass

Common sense tells us there are distinctions among the three main categories of travelers. Explorers seek out places where no one of their kind has been (though native guides familiar with the locations are generally indispensable to the 'discoveries'). Pilgrims and tourists travel not as explorers, that is to discover something new, but rather to experience what countless others have before them. Indeed, for both, the supporting apparati of travel – published guidebooks, popular travel accounts, professional trained guides, reliable means of transportation and often fully or reasonably equipped accommodations – are what sets this kind of travel apart from that of explorers. And so does something else: Except in fantasy and science fiction, explorers travel on a purely horizontal plane, pushing the edge of the frontier outward. Pilgrims and tourists often believe themselves to be traveling on a vertical plane, that is, visiting the past, or in the case of a visit to a more economically advanced society, the future.

Although the term explorer would seem to represent a distinct category in regard to other types of travelers, the truth is that all three share much in common: all value travel to out-of-the-way places that either do, or that purport to offer, unusual things to see or experience. For contemporary travelers with a substantial income and ready access to the world, these places should be, so to speak, on the bucket list. Their cache is enhanced if the locations are well beyond the reach of most either because of cost, commitment of time to reach them or willingness on the part of the visitor to endure sometimes unreliable and seemingly unsafe means of transportation (or at least that take some getting used to such as animal transport), the likelihood of having to eat exotic foods that challenge the non-native's sense of disgust, accepting sometimes primitive accommodations and facing danger either because of terrain, fauna or potentially hostile and aggressive inhabitants. Since many travel with the intention of writing about their experiences, the prospects of an arduous trip may be less a reason to go elsewhere than to welcome it as grist for the mill of a hoped-for spell-binding narrative. And in the modern era, that is, ever since the invention of this technology, most travelers expect to photograph or be photographed *in situ*, ideally next to something iconic as proof of having been someplace.

In regard to the temporal relationship among the three, we can assume that explorers precede tourists. But pilgrims are temporal floaters, coexistant sometimes with explorers and almost always with tourists. Moreover, certain locations lend themselves to a simultaneity of all three. In the latter part of the nineteenth and well into the twentieth century, travel to and within the Holy Land is a case in point: it had long been a place of pilgrimage, but since the time of Napoleon was becoming a popular site of exploration for would-be archaeologists and ultimately Bible enthusiasts. And it was increasingly attractive to European and American tourists. Although Muslim visitors visited Palestine on the way to Mecca as an *umra* or minor pilgrimage, tourism and exploration were more limited since religious authorities took umbrage at attempts to depict holy sites. Indeed, for non-Muslims well before the commercial publication of guidebooks, drawings often made furtively and inaccurately and, later, photographs of holy places undoubtedly laid the foundation for Western travel to the Near East.[1]

Guidebooks to the Holy Land

What brings these distinctions to mind is reading through nineteenth- and early twentieth-century guidebooks to the Holy Land. Baedeker's 1898 *Syria and Palestine* handbook itself takes issue with distinguishing the kind of excursions it purveys from leisure travel and immediately dismisses journeying to the region as a mere pleasure trip. There are, the book argues, much more colorful landscapes in Egypt than in the Holy Land and access to the few well-preserved ancient buildings such as Petra is difficult: 'the main object of a traveller to Palestine must be to call up the historical associations of the country, and in proportion as the traveller keeps this aim clearly in view and prepares himself for it, he will be able to overcome the inconveniences of the trip, the fatigue, the bad accommodation, and the monotony of tent life, and be preserved from disenchantment' (Baedeker 1898: xi). Rhetorically, the kind of travel it's purveying is more akin to what one expects from pilgrimage than tourism.

The emergence of the steam ship and the spread of railways across continents during the nineteenth century emancipated distant journeying from the sole possession of the truly intrepid. In the case of travel to Palestine, already by the 1880s one could proceed by train from Paris to Istanbul or journey for six days aboard a steamship from Marseilles to Alexandria and from there by French steamship to Beirut, with a boat leaving three times per month (Murray 1868: xlvii). From Egypt one could head overland to Lod or by boat to Jaffa, where travelers then boarded a train to Jerusalem or (some years later, after the city's founding) a coach to Tel Aviv. Actually, travel within Palestine offered various options. By the early twentieth century visitors could make the 53-mile journey from Jaffa to Jerusalem by train in 3 ½ hours, nine hours by carriage (though a shorter route), or for the truly adventurous and those determined to veer off the beaten path, over two days on horseback.[2] Once in the city, the would-be adventurer might find that he or she hadn't discovered anything at all. A typical travel handbook from that period refers the reader to a list of hotels then enumerates all basic services and addresses, including those of banks, tourist agencies, churches, hospitals, doctors, dentists and, it almost goes without saying, photographers.[3]

Explorers

There had long been guidebooks for how to travel to and within the Holy Land. But the latter part of the nineteenth century gave rise to so many, both hard-bound and softcover, that it's clear that we have entered the world of commercial tourism[4] with perhaps ten to twenty thousand visitors traveling to the Holy Land each year (Katz 1985: 52). The first systematic guidebooks appeared in the 1820s and 1830s, with major European travel guide companies producing consecutive issues shortly afterwards. The books included maps, illustrations, recommended sights and routes. Increasingly they included information on natural history, geography and archaeology based on recent findings, all indicating a secularized and scientific approach to interpreting the Holy Land (54–5).

Due to their suitability for winter excursions, by the 1870s Egypt and Palestine had become major components of Cooks' program. The British company pioneered group tours to the Holy Land by providing travelers with everything required after arrival (Friedheim 1992: 39). The growing success of these excursions to the eastern Mediterranean made it feel, as the owner of the company himself put it, like it was summer all year round (Brendon 1991: 129). Unlike the Riviera where visitors could easily make their own way unescorted, the eastern Mediterranean required planning and some discretion. The guidebooks warn that not all aspects of these journeys are for everyone. To counterbalance possible hardships and mishaps those who undertake them should have a purpose in mind. Cooks' 1872 volume for Egypt and Palestine warns that while many attempt to reach the Holy Land from Egypt, the journey is made across the desert on camels, and along sparsely traveled routes there is a possibility of encountering rapacious sheiks. Consequently, taking a boat to Jaffa and a bus from there to Jerusalem may be more advisable (Burns 1872: 63). The possibility of an encounter of that sort wasn't the only thrill travelers could anticipate, and many of these were precisely what lured travelers in the first place. Indeed, in Murray's 1868 *Handbook* the idea of exploration was more than just a rhetorical trope. The book opens with a listing of Arabic consonants and vowels and how they are written, and concludes with an 80-word glossary. The guidebook notes that although there has been a great deal of exploration and mapping for the region, there is still much to be done. Travelers who rely on a guidebook are admonished to 'ascertain by accurate astronomical observations the latitude and longitude of important towns and ancient sites... [T]he discovery of the true position of any prominent site would be an important addition to geography' (Murray 1868: xlix–xlviii). All inscriptions should be copied 'in whatever language previously unknown' (xlvii). For a tour or exploration into the interior, *Baedeker's* advises taking blotting paper that can be wetted and pressed on inscriptions. When dry, on it would be a copy of the impression. These can then be rolled up and kept in a botanist's canister (Baedeker 1906: xxii). *Murray's* instructs travelers to document ruins, especially those from Hauran to Palmyra. Would-be archaeologists among the travelers can excavate sites along the northern coast of today's Lebanon, and those with a leaning towards the visual arts should document the mosques of Syria, whose 'fretted minarets, inlaid walls, deeply-recessed doorways, marble courts, and arabesques interiors, are models of airy elegance – graceful and fantastic as an Arab poet's dream'. This encomium to Islamic architectural achievements is less than it seems because it's followed by the statement that, 'The best specimens are, like Mohammedanism itself, rapidly decaying' – a trope commonly embraced by Western travelers to the region (Murray 1868: xlix; Rogers 2011: 66–7).

The thrill of adventure is further elaborated in *Murray's*. Since Syrian roads are not always free of bandits, the guidebook advises that a revolver should be worn. In Palestine the peasants are generally armed yet rarely use their weapons against Europeans. There are times, however, when an escort savvy in local custom is helpful. If for some reason an attack seems imminent especially by Bedouin, 'no attempt at resistance should be made. Leave the matter entirely to your escort, and act as if you had no interest in it whatever. It will be well to explain to the enemy that you had no intention of breaking the laws of desert life. A calm and conciliatory bearing, aided in the end by a small present, will in nine cases out of ten clear away all difficulties' (xliii). Obviously, the warning was intended to assure visitors of their probable safety rather than to frighten them away, since the guidebooks' very existence depended on inducing more and more people to make the journey. Towards that end, *Murray's* proceeds to outline a series of customized tours. Recognizing that every traveler has in mind something in particular 'in making a "pilgrimage to Palestine"', the guidebook outlines four. To give a sense of the time commitment for some of these, let's take the example of the Grand Tour, which sets out in early February from Cairo, proceeds to Suez and Sinai then to Aqaba and Petra, to Hebron, Beersheba and the Dead Sea, finally arriving in mid-May in Beirut some 3 ½ months later! (xlix–L). The itinerary seems geared to those whose passion is the Hebrew Bible. For those devoted to the New Testament there is The Tour for Pilgrims, intended for the traveller 'who wishes to have his thoughts solemnized and his faith strengthened by a view of those scenes where the modest sacred events of our common Christianity were enacted' (Li).

Even with the use of a guidebook, travelers relied on local expertise, provided by a *dragoman*. The term is cognate in Semitic languages with translate (including the Hebrew *targum*) (Lewis 2004: 9). Conversant in various local and European languages, the dragoman served as a headman, without whom the logistics of a journey through the interior would be truly formidable. Perhaps because they were indispensable, dragomans had the reputation of considering their clients as little more than geese whom they felt duty bound to pluck (Brendan 1991: 121). Indeed, travelers were warned to use caution before engaging one. *Murray's* provides a page-and-a-half description of the dragoman one should book, explaining that the man arranges everything needed for the journey – tents, beds, animals, hotels and local guides. The guidebooks caution, however, that the route and timing should not be left to the dragoman's discretion but selected through the use of a guidebook. The traveler must insist that the dragoman follow that route, 'all difficulties and dangers notwithstanding', while warning the reader that stories about robbers are frequently fabricated by the dragoman to keep clients on his preferred path (Murray 1868: Liii).[5] Although many naïve travelers believe their dragoman to be the epitome of devotion, unknown to their clients, these seemingly loyal servants receive ten to twenty percent commission on every purchase. The dragoman, therefore, is unlikely to provide information on 'history, antiquities, statistics, or even places of interest out of the beaten track'. For those, the traveler must depend on his or her own reading (Murray 1868: Liii–Liv). Concerned about travelers' ability to select a reliable guide on their own, *Murray's Handbook* provides a recommendation and likely whereabouts in Cairo of a guide with whom the author personally made a 40-day journey (Murray 1868: Liv).

Karl Baedeker's 1906 *Palestine and Syria* also warns travelers to be wary about the information they are given about antiquities from guides or even dragomans. These have their own sense of what should be seen, and it may require some effort to convince a guide to deviate from his customary path. For tours that are longer than a few days, Baedeker recommends the use of a written contract with the dragoman, and provides a three-page single-spaced sample. On it should be the itinerary, explicit details about who defrays which costs, the means of transportation to be used and saddles if by horse, liability for injury to the animals and the right of clients to object if they feel the animals are being overburdened.

Camping

Although the 1898 *Baedeker* warns against the monotony of tent life, oddly enough the guidebooks paint an enviable picture of what awaited the outdoorsman.[6] This was not camping in the contemporary sense – a way to rough it, commune with nature and avoid the high cost of hotels. In regard to food and even to some extent accommodation, it was more akin to a cruise.

Cook's 1891 *Tourist's Handbook* has a two-page description of camp life, which assures prospective clients: 'When the camp arrangements are as they ought to be – and this is always guaranteed under the management of the dragomans engaged by Messrs. Cook & Son – camp life is delightful' (Cook & Son 1891: 7). The dragoman must see to it that there are tents for sleeping with mattresses, blankets, sheets, pillows and towel, a dining tent with table and chairs and arrange for a good cook and for guides and guards when necessary. The servants he hires 'shall be in every respect obedient and obliging, and shall be careful not to disturb the traveller's sleep' (Baedeker 1906: xviii).

Breakfasts are hardy and, of course, the utensils are cleaned and packed away, though not by the travelers. Mules carry everything – dishes, cutlery, tents and food and the alacrity of packing and unpacking is 'a continual source of amusement day by day' (Cook & Son 1891: 8). Freed from the drudgery of set up and cooking, travelers usually spend leisure time reading up on the places they're about to visit. Lunch is served at a convenient spot and consists of poultry, sardines, eggs, bread and cheese and two kinds of dessert. The length of the break for the noon-time meal often provides time for a nap. The day's travel ends at six or seven, by which time travelers can wash and unpack while watching the tents being pitched and dinner being prepared on a camp fire: 'The table of the saloon is generally gay with flowers gathered *en route*, and the general aspect of the social board is such as might be expected in the neighborhood of the Italian Lakes, but not in the wilds of Syria' (8).

Instructions to travelers in Karl Baedeker's *Palestine and Syria* are more detailed than *Murray's*.[7] Almost a full page is devoted to clothing: 'The tailor should be instructed to make the sewing extra strong, for repairs and the sewing on of buttons are dear in the East, not to speak of the difficulty of finding the tailor just when he is wanted' (Baedeker 1906: xxi). The guidebook suggests what items to bring, and also advises that 'valuable watches should be left at home' (xxii). An interesting entry has to do with lodging. Baedeker distinguished four categories with the so-called proper hotels available, of course, only on the major tourist routes. One can also find accommodation at hospices and convents, or in villages in which accommodation for payment is generally available at a notable's home. Peasant huts made of mud are to be used only as last resorts, since they swarm with fleas and vermin: 'The traveller should see that the straw matting which covers the floor is taken up and thoroughly beaten, and the whole place carefully swept and sprinkled with water. Every article of clothing and bedding belonging to the inmates should also be removed to another room' (xvii). 'By contrast, Bedouin accommodations are free of insects but they are lice ridden. Scorpions abound in Syria, though they tend to sting only if irritated. The traveller is safe from them if he/she sleeps on a raised bed' (xvii). Besides these tips, the handbook has some 17 pages devoted to Arabic, 11 pages to religion, 18 pages to history, five to art and includes a three-page bibliography.

The Bible

These guidebooks were intended for European and American visitors who were steeped in the personages and landscapes of the Bible. Their readers clearly wanted to experience the Holy Land's topography, flora and climate first hand, and information and itineraries are presented accordingly. Although the texts are generally not devoid of observations about the peoples, languages and religious practice of the local populations, their primary purview is hardly ethnography but the continued presence of the past and the remaining evidence within the landscape of both the Hebrew Bible and the New Testament. The opening sentence of *Murray's* 1868 *Handbook* states outright, 'the Bible is the best Handbook for Palestine; the present work is only intended to be a companion to it'.

By the latter half of the nineteenth century the main roads of Palestine were so well known that *Murray's* advises travelers not to bring navigational instruments such as sextant and compass. Aside from scientific observations there was little that the general traveler was likely to add. The same was not true for the artist or poet, who could delineate for everyone hallowed scenes from the Bible: 'Every nook and corner of Palestine ought to be made familiar to us, whether portrayed by its own bright sun, or by the pencil. Let artist and photographer continue their praiseworthy labors, till every hill and every vale, every proud column, and every prostrate wall, that has a story in it, is carried away to the far west'. Costumes, homes and utensils, weapons and tools for animal husbandry 'are all interesting, as all tend to throw fresh light on ancient history' (Murray 1868: xlvii).

Despite the significance of the Bible for the guidebooks, other sources were used as well, including Josephus's account of first-century Judaea, the Roman siege of Jerusalem and the consequences of the Great Revolt. Some guidebooks also included capsule summaries of the Holy Land's subsequent history, including the Islamic, Crusader and Ottoman conquests. Still, *Cook's Tourist's Handbook for Palestine and Syria* published in 1891 is quite explicit about the primacy of the Bible as a sourcebook. The book presumes that the typical visitor will be traveling by horse and sleeping in a tent. It was, therefore, printed in clear typeface to be read on horseback or by light in a tent. It also contains 'the full text of Scripture references, so as to avoid the inconvenience of having to turn to the passage in the Bible' (Cook & Son 1891: iii).

Who Went on the Tours?

Cook's tours to Palestine began in 1867, and within 5 years the company had escorted some two hundred people there. In 1879 John Cook could claim that three quarters of all American and British travelers to the Holy Land were handled by his agency. By 1882, that figure amounted to 5,000 visitors (Brendon 1991: 135). The growing numbers induced the company to purchase land outside the walls of Jerusalem for possible use as a hotel and central depot. Given what we know of the camping arrangements, it's pretty clear that the clientele for these tours was fairly well-to-do. Cook's assured travelers that they would be well cared for by a salaried staff of assistants. This included boatmen to meet ships in Alexandria. The staff would help them disembark, guide clients through the Custom House and escort them onto Nile steamers or onto transportation for the overland journey to Palestine. Similar service was available for those disembarking at Jaffa. The company also vetted its dragomans to secure only those vouched for by previous clients.

In 1881 John Cook himself spent time in Palestine and the result was a substantial remodeling of staff and equipment including tents and horses. Travelers had the option of hotel accommodations in major cities, but, apparently, most wanted to camp out (Cook & Son 1929: 5–7). For those preferring hotels, Cook's 1891 handbook indicates that the options are limited and a traveler is advised to write to the Cook's bureau in advance. He or she will be offered coupons at reasonable rates at select hotels. The traveler is thereby guaranteed satisfactory accommodation even in the busiest of seasons (Cook & Son 1891: 10), knows where to go upon arrival in an unfamiliar city and can find letters from home on his or her arrival. Most importantly, since charges are fixed and are known in advance, the traveler is able to calculate the costs of the trip before departure. The concern about cost suggests, I believe, the gradual degradation of travel from the purview of the upper to the middle class. It also suggests the increasing popularity of Holy Land touring in the latter part of the nineteenth century.

Cook's profited greatly from European development projects that mitigated the arduousness of traveling to and within the Middle East. By the 1920s the company was booking passengers on steamers with luxurious accommodation that went from Cairo up the Nile (Cook & Son 1929), and the construction of railroads and modern roads made travel to and within Palestine closer to what one would expect when journeying in much of Europe. Indeed, after World War I the 15-hour train trip from Cairo to Jerusalem included sleepers and a dining car, and there were now various sailings bringing passengers to Jaffa and Haifa. There were also plenty of options for trains carrying passengers from England to Istanbul. One could even book passage on an plane between London and Alexandria and Baghdad. The trip took four days, though passengers weren't in the air for much of that. If Cook's own promotional brochure can be trusted, even hotel accommodation had improved substantially, with establishments in Jerusalem and Tiberias offering some rooms with private bath. A company brochure boasted that a fully equipped modern hotel was scheduled to open in Jerusalem in 1930 (Cook & Son 1929: 67–8).

It's difficult to say who the clients were for these trips. Of the approximate 600 distinguished names of the company's patrons listed at the back of a brochure announcing Cook's 1881–2 Programmes and Itineraries to Palestine, none are what could readily be identified as Jewish. Quite a few are Christian clergy and their wives, military officers and government officials and their spouses. Even European royalty made their way onto the list. Why this absence? Did Jews not go on these tours or was it simply that they were not distinguished enough to be included in the promotional brochure? Of course, when Cooks purveyed the Holy Land it wasn't the Jewish Holy Land the company had in mind. Moreover, the truly religious of any denomination were not Cook's preferred clientele. In 1881 the company received a commission to take a thousand French pilgrims from Marseilles to Jerusalem and back – a party consisting of 356 Jesuit priests plus lay persons. Blaming the Jesuits' handling of the agreement, John Cook vowed he would never again mix business and religion. Instead, the company pursued a moneyed class of travelers. Some 13 years after the disastrous French pilgrimage, Cooks escorted the Prince of Wales' two sons to Palestine and then in 1898 the German Kaiser (Brendon 1991: 138–40).

The absence of Jewish names on the list of Cook's patrons may have had something to do with the fact that the company's version of the Holy Land was not how Jews would want to experience their ancient homeland. Interestingly, the same could be said for some Christian denominations. Catholic and Orthodox visitors had a large array of churches and shrines, with many selling souvenirs, the acquisition of which was an integral part of the journey. Some Protestants were repulsed by the ornate churches with their trinket vendors. To them the churches bordered on idolatry and the trinket vendors were a reenactment of the money changers in the Temple. They responded by developing their own sites and pilgrimages and by seeking 'Christ in the outdoors, in the wild untamed wilderness area of Palestine. Since these places were pristine, they were thought to link directly to biblical times' (Rogers 2011: 145–6).

But even ostensibly non-religious, or better put, religiously neutral sites that all visitors would see were less exhilarating to Jews than to others, especially when seen in the company of those for whom the ruins have a distinctly Christian allegorical significance. For them, the destruction of the Temple and the exile of Jews from Judea were signs of Divine punishment for rejecting Christ. Believing Jews, too, could see in the ruins an allegorical significance – Jeremiah's prophecy of destruction based on collective sin intertwined with the prophesies of rebirth and Redemption. But certainly, for Jews, besides the thrill of treading on ancestral terrain, the reality of what they saw was hardly uplifting.

Jerusalem had been a ruin ever since the last revolt against Rome in 132–5 CE and the subsequent Roman reconstruction of the city transformed it from a Jewish to a pagan center. Some 1,800 years later there was hardly a trace of the Judaean capital's first-century majesty – once a wonder of the ancient world. Cook's 1907 *Handbook* describes 'the burden of the Old Testament' quoting Lamentations, 'Zion spreads forth her hands, and there is none to comfort her' (Hanauer and Masterman 1907: 61). The Handbook describes modern Jerusalem as a disappointment and 'much less than the imagination had pictured', noting that 'the city whose streets Jesus trod had once been about a third larger. Then Zion, a large part of which is now a ploughing field, was covered with palaces; and on every side, where now the husbandsman pursues his toil, or desolation reigns, were magnificent structures befitting a great capital' (61). The guidebook proceeds by remarking how little remains of ancient Jerusalem, noting how even its walls date only to the sixteenth century. Although a modern Jewish community existed in Jerusalem, for a tourist with little knowledge of, or interest in, the panoply of Jewish cultures and religious practice, of which there was plenty to observe in Jerusalem, there was nothing of interest to see among local Jews. The word 'see' is critical here because the primary sense evoked by tourism is the visual, and unlike smell or taste, it is often the sense that is the least offended by encounters in strange locations. *Cook's* earlier 1891 edition describes the Jewish Quarter's synagogues as 'used respectively by the Sephardim and Ashkenazim and are singularly devoid of interest' (163). For those who travel to see the sites and to enjoy the visual splendors of the ancient world, Jewish Palestine had nothing to offer. It was a ruin and much of that had yet to be excavated.

For Christians, the connection to the Holy Land had everything to do with a cult of hagiography whose figures were spawned in historical times and whose lives and final resting places often enough are associated (albeit often without solid foundation) with specific locations. Designated for purposes of veneration, tombs and the churches built upon them, these sites draw visitors who expect certain benefits through such contact. A steady stream of visitors enters Jerusalem's Church of the Holy Sepulcher, many kneeling at and kissing the site where Jesus' crucified body was lain. Some place bags full of commercially made olive wood souvenirs and bottles of Jordan river water on the slab so that they might absorb the holiness and make the souvenir less a trinket than a relic to be given as gifts to family and friends upon returning home.

Inanimate sites can have charisma. As the putative burial place of the Christian god, the Church of the Holy Sepulcher is unique in its magnetism, which is further enhanced by legends attached to it.[8] Except for the tombs of famous rabbis scattered around the world, there is nothing comparable in Judaism. King David's tomb was readily discredited by one guide I talked to as containing no past Jewish king but rather, to quote him, the body of an anti-Semite – a knight Crusader. Even the foundational figure of Judaism, Moses, has no known or even presumed burial location. Although tombs of the patriarchs are putatively designated, the same people are venerated by Moslems who erected mosques over them and traditionally forced Jews to remain at the perimeter. Some biblical prophets and kings are associated with certain tombs and caves in and around Jerusalem or elsewhere in the country, but the sheer modesty of the tombs, if one can call them that, and their somewhat obscure status in the pantheon of 'saints' would make them of interest only to a true aficionado of the Hebrew Bible. Without a guide the sites would typically be overlooked, and for good reason; they're not monumental and don't provide that much to see.[9] The most important Jewish religious structure in Judea had been destroyed by Roman siege and was appropriated through successive conquests for the religious activities of the city's conquerors. The rabbis debated whether the departure of the Divine Spirit after the Temple's destruction has made grounds once forbidden to all but the High Priest once a year, no longer sanctified. If so, there is no longer any danger of defiling sacred ground. But since the issue is in dispute, observant Jews traditionally shied away from visiting anything other than the Western Wall – at some distance from the Temple itself and certainly from the Holy of Holies.[10]

Although there remained an attachment to some historical Jewish sites akin to the saint veneration of Christians – the Temple Mount's Western Wall or Rachel's Tomb, for example (the latter was traditionally popular among Jews and was designated as such by Ottoman and Mandate authorities) – the only thing comparable to the acquisition of religious trinkets for Jews would probably have been the purchase of postcards of drawings and then of photographs of these sites. Although there was no need to travel to acquire these mechanically reproduced items, by the early twentieth century, art and handicraft began to be produced on Jewish themes within the Bezalel Museum in Jerusalem and at least some of the purchasers of this newly minted Judaica were foreign Jewish visitors.[11] Regarding saint veneration among Jews, there are, to be sure, the tombs of famous rabbis both within and outside of the Holy Land. But there was never exactly a cult of relics in any way equivalent to that of Christians.

In his travel book, *Innocents Abroad*, Mark Twain was interested in Old World tourist sites, and he had great fun with the remarkable proliferation of Christian reliquaries and their sacred contents. When passing by the ancient residence in Jerusalem of St. Veronica who wiped the perspiring brow of Jesus as he labored under the heavy weight of the cross he bore and whose face then appeared on the kerchief, Twain remarked, 'We knew this because we saw this handkerchief in a cathedral in Paris, in another in Spain and in two others in Italy. In the Milan cathedral, it costs five francs to see it, and at St. Peter's, at Rome, it is almost impossible to see at any price. No tradition is so amply verified as this of St. Veronica and her handkerchief' (Twain 2007: 446). Despite the sarcasm of Twain's comment, the observation suggests that the replication of religious objects purveyed as relics does nothing to demystify them. Instead it constitutes for the multitudes who believe in the efficacy of these objects' links to the Divine a kind of corroboration. For a wit like Twain's, the flim flam of the claims was grist for the mill.[12]

Crowded out by site or saint veneration of others, inclined to see them as abominations, and sometimes skeptical of the historical veracity of sites claiming to be tombs of biblical figures,[13] Jews connected to the Land of Israel through produce used or consumed during certain holy days, through the repeated recitation of the liturgy in which the Land of Israel is a central motif, and in Judaism's eschatology. In earlier times a particularly devout Jew might travel or wish to travel to the Holy Land towards the end of life, anticipating burial close to one's biblical ancestors and shortening the journey he or she would undergo after the revival of the dead at the End of Days when all Jews will assemble in and around Jerusalem. Some still insist on burial there – an easier request to fulfil today *post mortem* through airplanes and refrigeration. Still, unlike Christians, Jews had neither a well-marked nor a particularly extensive pilgrimage route within the Holy Land. Even the most significant monument on that route was no more than an ancient retaining wall abutting a narrow courtyard. Unlike many Christian denominations, Jews had no religious orders whose mission it was to house and care for those who made the journey. Nor would Jews have felt comfortable or even welcome in these institutions. Indeed, Jews were not welcome, except by government fiat, in many of the prime pilgrimage sites throughout the Holy Land.

Take the following account of Sh. Erdberg, an East European Jewish immigrant to the United States who visited the Holy Land in the 1920s. Wandering along the Via Dolorosa near the Fourth Station of the Cross he suddenly heard the distinctive melody of Talmud study. To his amazement, he had discovered a yeshiva ensconced between the Fourth and Fifth Stations. Apparently, a Jew had purchased the property and given the space to a yeshiva. The 'invasive' Jewish institution fended off the protests of local Catholics partly with the help of the Russian Orthodox Church delighted to stick a needle in the eye of their Roman adversary.

Elsewhere in his meanderings in Jerusalem the author reflects on the fact that despite being raised in Eastern Europe and having lived in the United States, he himself had never experienced a pogrom nor had he ever seen a lynching. In Jerusalem, he managed to survive both. These occurred when he wandered into the Church of the Holy Sepulcher. Disoriented, in the dark and unfamiliar interior, Erdberg spotted a Christian Arab and promised to reimburse the man handsomely if he would act as his guide. The man lit a candle then took the author downstairs to visit the cellar where Jesus was held prior to his Crucifixion and then to see where the crucified body lay. 'I don't know how it happened, but somehow I blurted out the question: "You say, however, that he ascended to heaven. So how come you say that he's buried here?"' In asking the question Erdberg revealed the fact that he was a Jew. A commotion erupted, with people shouting 'Jew, Jew, Jew' in Arabic, their eyes as fiery as pogromtshiks. The author whipped out his American passport as a shield and suggested that if the crowd had any issues with him they could settle things at the American consul. Fleeing the church, Erdberg swore off seeing any other Christian buildings in the Holy Land. Some distance away and able to breathe easier, he spotted an Arab running towards him. Fearing another fanatic seeking vengeance, he quickly realized that the 'fanatic' was his guide and that Erberg had bolted from the church without paying for the man's service. The crisis over, the author lamented the Jewish fate: 'They don't permit Jews to cross the threshold of the "tomb". It's *their* holy of holies. O.K. But why are our holy places in strangers' hands? Our holy places were taken by others, and kept by them. Their holy places they protect, they won't even let us lest we defile them!' (Erdberg 1926: 2–3).

Erdberg's consolation was the Russian church. At one time, no Jew would dare to enter it for fear of being beaten by the priests. Today they're still hostile, but the nuns not so much. They're grateful for any contribution no matter how small. Since the Revolution, they've been abandoned, are quite poor and in dire need of contributions by tourists (74).

The Jewish Visitor

By the latter part of the nineteenth century, Palestine represented a different kind of Jewish revival – a secular rather than religious one. And that revival attracted Jewish settlers who removed themselves spatially from the older communities of religious Jews in Jerusalem, Hevron and Tsefat. Many also came as visitors, or as writers and journalists, to examine the accomplishments and continuing struggle of those who had settled in newly purchased land. This was a period when Jews, and especially East

European Jews, were on the move. They were interested in other places and that interest was more than idle curiosity. Increasingly, with Europe's new political configurations after World War I, the position of Jews within some of these states proved fragile enough to make emigration if not an immediate necessity then at least a strong consideration.

To be sure, Jews who traveled to Palestine were also interested in seeing the sites that pertain to ancient Jewish history. But they were not always pleased by what they saw. As one traveler noted, Jerusalem was impressive only when seen from a distance (Hoffman 1923: 53). Up close it was poor. Its streets were narrow and dark and the Old City market through which one passed on the way to the Western Wall was filled with tiny stalls selling food items covered with flies. It was the Orient in the most negative sense of the term. And the Jews one saw wore an agglomeration of colorful but odd-looking costumes. They mostly spoke Yiddish and the Sephardic equivalent, Ladino.

Yehoash, the poet and translator into Yiddish of the Hebrew Bible journeyed from New York to the new agricultural colony not far from Tel Aviv, Rehovot. Although, he intended to settle permanently the outbreak of WWI forced him to return to the United States. While living in Rehovot he travelled to some sites of biblical significance. Yehoash's accounts of these have a unique quality because of his fascination with language and his tendency to interweave legends with his own observations of places and people. He was anything but entranced by the ruins of ancient Israel, and in support of an initial hesitation to visit Jerusalem he suggested something that only a Jew deeply immersed in the Hebrew Bible could articulate: 'there is more of the past in a single chapter of the *Book of Kings* than in all the stones of Jerusalem' (Yehoash 1897: 149). When he finally makes a very brief visit to Jerusalem, he navigates narrow streets and a dark store-lined passageway, descending until coming to a narrow path of stairs. Before him is the Western Wall with the characteristic chiseled massive stones of the Second Temple era. Since it was early morning, only a half dozen of the usual beggars were in place. He saw an old Jew lying on a stone and weeping over a book of psalms, an old woman reciting mechanically from a book of women's Yiddish prayers and another woman mumbling prayers in Ladino while kissing one of the wall's stones. The scene did not move Yehoash in any religious sense. What he was left with, rather, was a sense of bitterness and despair of seeing a desperate people that had come to pound on its last entryway (155).

Yehoash completes this narrative with a legend about a newly appointed sultan in Jerusalem from several hundred years ago. Every morning peering out of his window the sultan could see an old woman struggling with a large basket of waste, which she would dump on a mound not far from the palace. The sultan ordered his guards to bring the woman to the palace and he then asked how she dare do such a thing. The old woman replied: 'Sire, it's not my fault. We were ordered to do so from generation to generation. Titus brought my forebears from Rome and ordered that they, their children and grandchildren should throw waste on the site of the Temple. So, we obey the command until this day.' The Sultan ordered the mound excavated and after months of labor the Western Wall was uncovered. When it was cleaned, the Sultan assembled the Jewish leaders of the city and told them to rebuild the Temple in all its former glory. The leaders thanked him but asked that the Western Wall remain as is. Their only request was to be able to come to it every Friday evening and weep, as is their custom. To rebuild the Temple is forbidden until the coming of the Messiah. Concluding the visit, Yehoash headed back to Rehovot and wrote, 'The stone wall stood heavily on my mind and all the more eager was I to return to the lovely verdant Rehovot. All the faster did I want to discard the cold dead ruins and once again breathe in the fresh air of the vineyards' (156–7).

Of course, what might evoke feelings of disgust encountered in the flesh could always be extrapolated and aestheticized as sketches[14] or photographic reproductions. Images of the Western Wall or Rachel's Tomb and other monuments received a Zionist interpretation, retaining their value as symbols of redemption but converted to nationalist values and dreams of renewal and the return to the ancestral homeland (Cohen-Hattab and Kohn 2017: 70). Another welcome relief to the strictures of the past were the Jewish agricultural colonies that had sprouted up, and, of course, the emerging Jewish metropolis of Tel Aviv with its Hebrew speakers, Jewish policemen and bus drivers, Jewish industries and seaside hotels. Small though it might have been earlier in the twentieth century, like the agricultural colonies, the new city represented a site less constricted by the Bible and the decrepitude of the Orient and the holy sites of Jerusalem, where every buried body included a living outstretched hand demanding alms, or a previously unseen presence within dark tombs appearing out of nowhere to recite a prayer for the visitor's wellbeing and demanding payment in return (Hoffman 1923: 67–8). Tel Aviv was free of the past and it allowed visitors to look forward to what an autonomous Jewish polity might become.[15]

Let me return for a moment to the initial paradigm of explorers, pilgrims and tourists. I would like to suggest at least one more category – writers. These are people whose purpose in traveling is to convey to a broader audience the place visited and the encounters experienced. Very often, the accounts are forwarded to the newspapers for which they write and appear in serialized form while their authors are still traveling. Consequently, the accounts tend to be episodic, though some do provide interesting observations, analyses of situations and people[16] or, as in the case of Yehoash, may include narrative gleaned from various sources unlikely to be gathered by a casual visitor. Let me give one more example, again from Yehoash. Visiting Hebron, he is shown a hole in the gate of the Jewish ghetto. His guide, a local Jewish dignitary, explains that many years ago an Arab pasha ruled Hebron who hated Jews. One day he imposed an enormous tax on the community and, if not paid by a certain date, he would subject its members to violence. The community fasted and prayed. In the middle of the final night before the tax was due a stranger pounded on the gate but the guard refused the man entry. The stranger then struck the gate with such force that it bore a hole through it. His hand reached through the aperture and presented the guard with a cloth package. A voice from the other side of the gate instructed the watchman to give this to the pasha. The hand promptly disappeared. The next day the community elders presented the package to the pasha. When he examined the contents his face turned white. The pasha revealed that the night before Abraham had appeared before him and demanded all the jewelry of his most beloved wife. He did as requested and now before him wrapped in the cloth was the jewelry. The pasha begged forgiveness and from then on treated Hebron's Jews well (Yehoash 1897: 64–6).

Although Yehoash's fascination with language and narrative makes his account somewhat unique, there is often some degree of ethnographic and even linguistic material in the travel narratives of other European and American Jewish visitors to Palestine during the first part of the twentieth century. If the guidebooks reviewed above are any indication of how many well-to-do people traveled to, and within, the Holy Land, the Yiddish accounts suggest how different such travel was for many of the Jews attracted there. Of course, Jews had access to a local population who spoke their language – whether Yiddish, Russian or sometimes Hebrew – so there was often no need of professional guides and interpreters. And there were enough local Jews whose primary language was Arabic to help guide those traveling to and within sites of Jewish interest located within Arab communities. But these Jewish travelers were not really explorers. And aside from visiting sites of Jewish interest they stuck pretty much to the beaten path. Indeed, none of the accounts I've looked at suggest the kind of penchant for exploration and discovery one finds outlined in the non-Jewish guidebooks. Rather than encountering the past, Jewish visitors brought with them a past they knew well. Since they relived the Jewish Holy Land in the weekly Torah portion, they didn't need to see the ancient ruins to bring that world back to life. Moreover, since ancient Jewish sites like Jerusalem had been appropriated by Christian missionaries, if they could not circumvent those sites they could at least ignore the parts of them that were offensive or of no interest to Jews.[17]

A Yiddish Handbook

In 1908/9, A. Grajevsky, associated with Tabor travel bureau of Jerusalem, published a small handbook in Yiddish for travelers to and in the Land of Israel. The introduction stresses the fact that in none of the many international handbooks are the specific needs of Jewish travelers addressed – a lack the author notes is frequently mentioned by clients entering his travel bureau (Grayevsky 1908: 2). Like a typical handbook, this one speaks to the practical, including what to bring – dark glasses, binoculars, a flask, a small Hebrew Bible, reading material, especially Davis Treitsch's *Eretz Yisroel*, and books on Rishon leTzion, books by Jewish travelers from the past and more recent ones to Jerusalem. Treitsch's handbook that Grajevsky refers to is more an inventory of Jewish settlements and developments within Palestine than a guidebook, since it lacks the typical set of itineraries and the plethora of historical information. Its primary focus is geography and agronomy, especially the Jewish agricultural colonies that by then dotted the landscape of Palestine. But it does include a great deal of practical information, including postal routes and offices throughout the region but also in the agricultural colonies, telegraph charges, train lines to and within Palestine, ports, import duties, security, doctors, pharmacies, hospitals and schools. It includes a 15-page inventory of the Jewish agricultural colonies, preceded by a sketchy history of Jewish travelers to, and settlers in, the Holy Land since the twelfth century. This is followed by a more detailed history of settlements since the first part of the nineteenth century (Trietsch 1910: 155–68). The section is followed by a few pages on the German agricultural colonies dating from the second half of the nineteenth century (168–70). The Trietsch handbook has a technical quality to it, providing basic information and wherever possible statistics, especially about Palestine's emerging economy.

Compared to that of Treitsch, Grayevsky's handbook reads more like the commercial travel handbooks reviewed above, though obviously with much more information of specific interest to the Jewish traveler. For those routed through Vienna, for example, Gayevsky provides information on hotels, kosher restaurants, cantors, but also more general information on travel agencies, mail and telegraph, tobacco and exchange rates of Turkish coins to other currencies. Grayevsky leads his Jewish travelers by the hand, with practical advice on travel documents, especially those that need to be secured for Turkey, and where the train station is located in Budapest that takes travelers to Constantinople (Grayevsky 1908: 31–2). Once in that city the guidebook is explicit about the four hotels that cater to Europeans, with one standing out for quality, the reasonable price of its food and lodging and the friendliness of the owner, who is always eager to help a foreign visitor (31). Grayevsky provides information for travelers heading from Constantinople to Damascus and Baalbek via Beirut. Like elsewhere in the handbook, the information is very concrete – the price of tickets, the schedule of trains going and coming and the length of the journey (40–1). The author includes a section labeled history and indicates the strong connection Damascus had with the Bible's book of Kings. Another paragraph discusses the Jewish community, which the author describes as small and diminishing, consisting only of Sephardic Jews, 'of a special type and with a strong temperament' (43).

The handbook outlines various ways to reach Palestine, including boat to Haifa (not recommended during the rainy season since land travel heading south is treacherous) or Jaffa (sometimes difficult to disembark due to stormy seas). It then lists hotels in Jaffa, and where to purchase alcohol especially Carmel wines, places to pray, doctors, pharmacists, tobacco vendors, banks, newspapers and the information bureau, travel between Jaffa and Jerusalem and, finally, hotels in Jerusalem. Regarding itineraries, the handbook divides Jewish Palestine into two (non-Jewish Palestine is almost entirely ignored): Jaffa is the heart of the agricultural colonies while Jerusalem is the head of the Land of Israel. Jaffa is the key to the new *yishuv* (prestate Jewish community in Palestine) while Jerusalem is its soul. So travelers should divide their time as follows: (1) Jaffa with the surrounding colonies in Yehuda; (2) Jerusalem with the historical new and old surroundings; (3) Haifa including the Galil, Shomron and the colonies (43). The waters at the port at Jaffa can be rough but the Arabs who maneuver the small boats used in disembarking are very skillful. (The comment is particularly interesting in light of the soon to emerge tension between the growing Yishuv and Arab nationalists, none of which is apparent in Grayevsky's handbook.) The author reminds us that Jonah had headed to this port when he fled his calling as a prophet in Tarshish, and it was in this port that King Hiram brought the cedars of Lebanon for Solomon's Temple. Jaffa has a large number of Jewish hotels, and the guidebook's author recommends Hotel Bella Vista built for foreign travelers with the latest of amenities including large airy rooms not far from the sea. The kitchen is up to European standards and strictly kosher, bathing facilities are available with fresh or salt water and each afternoon an orchestra plays on the grand terrace.

Just how different this handbook is from those for non-Jewish Holy Land visitors with their trope of exploration is readily apparent by the focus on the kinds of comforts that relatively well-off Jewish travelers might want: the author includes a small section on alcohol and recommends the (one assumes kosher) wines and liquors of Rishon l'Tzion. The hotel offers both foreign and local beers plus mineral water. Synagogues are in abundance. The hotel itself has a place for billiards and dominoes. Jaffa also has a Jewish club, and though entry is restricted by subscription, members can bring in guests. The guidebook also notes the existence of a library, and lists the names and locations of various doctors, banks, magazines, newspapers, and the location of the information bureau.

The handbook then describes in great detail the train route from Jaffa to Jerusalem and takes care to note both the current Jewish and German Templar agricultural colonies along the way as well as the sites connected to the Hebrew Bible, the first-century Jewish revolt and mishnaic-era Jewish events and personages (Grayevsky 1908: 57–8). In Jerusalem the narrative continues with a brief description of the main sites of the city and then a detailed tour of the Temple area, with its two mosques, and some historical information on their construction plus details on their interior architecture and design and, of course, the Foundational Rock believed to be the altar upon which Isaac was to be sacrificed by Abraham (some Christians believe this to have been the site of the Crucifixion as well). The description includes both Jewish and Muslim traditions about the site, including Muslim belief in the significance of the Temple Mount for the Day of Judgement and Mohammed's night flight to heaven. The account itself might easily have been drawn from any one of a number of travel handbooks. But there are details here and there that are unique. In the description of the Western Wall, for example, the author mentions the little notes inserted by those going abroad and seeking assurance of a safe return, or tourists who break off pieces of stone to take away as souvenirs. And growing on the wall are various sorts of grasses which visitors tear off. These are later cooked and fed together with a broth to the sick (88).

Subsequent descriptions of the city and recommended tour routes pay almost exclusive attention to Jewish dwellings, businesses and religious institutions of various traditions based on places of origin (94). All of the excursions the handbook outlines are short, ending back at the hotel in Jerusalem. No overnighters, no camping and only occasionally resorting to horse travel where necessary but more typically motor coach. Grayevsky mentions an intended second volume, but I have found no trace of one. Although he makes reference to the agricultural colonies in the first volume, there are no excursions to them. Perhaps they were to appear in the second volume. One point seems clear: Grayevsky's handbook was not intended for an intrepid traveler. For them a slew of other guides already existed in a variety of languages, many of which would have been accessible and comprehensible to the typically polyglot Jewish traveler. But Grayevsky's handbook suggests that by the time it was produced there were any number of travelers who really had no interest in exploring or discovering anything for themselves. They wanted to see sites well known to them from the Hebrew Bible or the Mishna. The more spectacular tourist sites of Christianity and Islam were of no interest to them. Like everyone else, including the various denominations of Christian visitors, they traveled to the Holy Land with blinders on, 'seeking', in the words of Mark Twain, 'evidences in support of their particular creed... [T]hey had already made up their minds to find no other, though possibly they did not know it, being blinded by their zeal' (2007: 393). Whether tourism, pilgrimage or exploration, what is interesting about travel isn't just the intention of the journey or what the visitor is shown in packaged tours. It is as Mark Twain suggests, what he or she sees but doesn't see, blinded by zeal. How much better this world would be, especially in the lands of the Bible, if only this blindness were just a matter of travel.

Notes

1. For an interesting study of travel to Palestine prior to the emergence of the commercial guidebooks, see Ben-Arieh 1979.
2. *Guide to Palestine and Syria* (London: Macmillan & Co., 1908), 11–18.
3. Ibid., 18.
4. One study of American Protestant travel to Palestine counts some 500 pilgrimage narratives published between 1850 and 1917 (Rogers 2011: 22).
5. This is pretty much Mark Twain's assessment, though he does mention sham attacks perpetrated by Bedouins hired for this purpose by the pilgrims' Arab guards. After the battle, the Arabs and Bedouins had lunch together, divided the baksheesh and escorted the group to the city (Twain 2007: 460).
6. In a promotional brochure for its tours, Cooks referred to this type of travel as 'a sealed book except to a comparatively few wealthy noblemen and distinguished students' (Cook & Son 1881: 5).
7. The 1898 edition is probably the third English edition and is based on the fourth German edition.
8. In his travel narrative to the Holy Land, Mark Twain describes the Church of the Holy Sepulcher and the spot where the mother of the Emperor Constantine found the three crosses of the crucifixion several hundred years after Jesus'

death. The story attributed to the discovery relates to which of the three was used to crucify Jesus. A priest devised a plan to present each of the three to a noble lady who lay very ill in Jerusalem. Each in turn sent the lady into a swoon from which she recovered with her health fully restored only after the presentation of the third and true cross (Twain 2007: 434–5).

9. In his visit to the Land of Israel, Pincus Puchkoff describes spending a day visiting graves in Jerusalem. His guide was a young Jewish man born in Jerusalem and well versed in the Hebrew Bible and the Talmud. The man, according to Puchkoff, was steeped in Torah, history, archaeology, theology and whatever else one might want. One of the first sites he showed the two couples who were traveling together was the newly excavated fourth wall of the city. The discovery proves definitively that the putative tomb of Jesus cannot be Jesus' resting place since it lies between the third and fourth walls of the city, and Jewish custom at the time was to bury the dead outside the walls of the city (Puchkoff 1928: 96–7).

10. In recent decades, some religious nationalist Jews have begun to insist that prayer on the Temple Mount is a necessary prelude to the final Redemption and the return to Temple worship.

11. The museum had both an art school and a gift shop (Puchkoff 1928: 91).

12. Jeffrey Melton argues that what runs through Twain's writings on the Old World is the frequent sense of disappointment that the places visited pale in size, often by their obvious tackiness in comparison with how they were first encountered in illustrations or descriptions and imagined from afar. It is the tourists' faith that what he or she sees is real despite the obvious contrivances that prevents this reality from turning into skepticism. Moreover, it is memory that reshapes touristic experience, making even unpleasant ones palatable, and memorable in a positive sense (2002: 76).

13. After visiting the Western Wall, Abe Cahan expresses the possibility of doubt as to whether or not the wall has any connection to the ancient Temple and argues that such a question is not unwarranted. Visitors to Palestine are shown hundreds of remains from biblical times, the majority of which have no historical basis in fact. He includes in this list Rachel's Tomb, Elijah the Prophet's cave and the Tomb of the Patriarchs: 'Not only is there no scientific proof that these sites are what people claim them to be, but there is, indeed, scientific indication that they are not'. Still, he concludes that regarding the Western Wall there is no doubt that it was a part of the structure that surrounded the Temple (Cahan 1934: 41–2).

14. The artist Raskin's book is based on two journeys he made from the United States to Palestine in the 1920s. The book includes a number of his drawings, but it is fundamentally a work of travel reportage (Raskin 1920).

15. Although unusual, not all accounts were optimistic about that future. B. Hoffman visited Palestine in 1923, first traveling through Europe. His observations were particularly negative, convinced that there was no viable economy in Palestine to sustain an increase in the Jewish population. The primary business he saw was sending request letters for contributions abroad and selling soil for Jewish

burials. The Mandate Palestine is half the size of Switzerland. The latter has a population of 3 ½ million, so Hoffman calculated that under the best of circumstances i.e. emulating Switzerland's agricultural fruitfulness, Palestine couldn't support more than 1,900,000 people. Given the presence of some 700,000 Arabs already, a sizable increase in the country's Jewish population is neither practical nor viable without some sort of economy (Hoffman 1923: 156, 161–2).

16. The journalist and editor of the *Forverts* visited Palestine twice in the second half of the 1920s. Among others, his observations about local Arab dress and racial types as well as the language facility of Jerusalem Jews are particularly interesting (Cahan 1934).

17. As Abe Cahan writes, reading the various pamphlets given out by travel bureaus, one would never know that Jerusalem had anything at all to do with Jews. Instead all one learns is the city's association with Jesus and the apostles: 'He stood here, he went there, here this happened to him, there something else happened to him' (1934: 29).

Bibliography

Baedeker, K. (1898), *Palestine and Syria Handbook for Travellers*, Leipsic: Kara Baedeker.

Baedecker, K. (1906), *Palestine and Syria with the Chief Routes through Mesopotamia and Babylonia*, Leipzig: Karl Baedecker.

Ben-Arieh, Y. (1979). *The Rediscovery of the Holy Land in the Nineteenth Century*, Detroit: Wayne State University Press.

Brendon, P. (1991), *Thomas Cook: 150 Years of Popular Tourism*, London: Secker & Warburg.

Burns, J. (1872), *Help-Book for Travellers to the East*, London: Thomas Cook.

Cahan, A. (1934), *Palestine: a bazukh in yor 1925 un in 1929*, New York: Forverts.

Cohen-Hattab, K. and A. Kohn (2017), 'The Nationalization of Holy Sites: Yishuv-Era Visual Representations of the Western Wall and Rachel's Tomb', *The Jewish Quarterly Review*, 107 (1): 66–89.

Cook, T. & Son (1881), *Programmes and Itineraries of Cook's Arrangements for Palestine Tours*, London: Thomas Cook & Son.

Cook, T. & Son (1891), *Cook's Tourist's Handbook for Palestine and Syria*, London: Thomas Cook & Son.

Cook, T. & Son (1929), *Programme of Arrangements for visiting Egypt the Nile Sudan Palestine and Syria*, London: Thomas Cook & Son.

Erdberg, S. (1926). *For a Id nokh Eretz Yiśroel: rayze bilder*. New York

Friedheim, E. (1992), *Travel Agents: From Caravans and Clippers to the Concorde*, New York: Travel Agent Magazine Books.

Grayevsky, A. (1908), *Rayze handbukh Eretz Yisroel*, Jerusalem: Tabor Travel Agency.

Hanauer J. E. and E. W. Masterman (1907), *The Cook's Handbook for Palestine and Syria*. London: Thomas Cook & Son.

Hoffman, B. (1923), *Mayn rayze in Eretz Ysroel*, Warsaw: Di velt.

Katz, S. (1985), 'The Israeli Teacher-Guide: The Emergence and Perpetuation of a Role', *Annals of Tourism Research* 12 (1): 49–72.

Lewis, B. (2004), *From Babel to Dragomans: Interpreting the Middle East*, New York: Oxford University Press.
Melton, J. A. (2002), *Mark Twain, Travel Books, and Tourism: The Tide of a Great Popular Movement*, Tuscaloosa: University of Alabama Press.
Murray, J. (1868), *A Handbook for Travellers in Syria and Palestine*, London: John Murray.
Puchkoff, P. (1928), *Mayne ayndrike fun Erets Yisroel*, New York: Puchkoff.
Raskin, S. (1920), *Erts Yisroel in vort un bild, 1921–1924: ayndrukn fun tsvey rayzes*, New York: Reznik, Menshel & Co.
Rogers, S. S. (2011), *Inventing the Holy Land: American Protestant Pilgrimage to Palestine, 1865–1941*, New York: Lexington Books.
Trietsch, D. (1910), *Palaestina Handbuch*, Berlin: Orient-Verlag.
Twain, M. (2007 [1869]), *The Innocents Abroad*, New York: Signet Classics.
Yehoash (1897), *Fun New York biz Rehovot un tsurik*, New York: Hebrew Publishing Co.

Chapter 3

BACK TO THE GARDEN:
BRINGING VISITORS TO AMERICAN EDENS,
1885–1956

Brook Wilensky-Lanford

Despite the use of the word 'Eden' or 'paradise' in adverts for innumerable tourist destinations worldwide, the correlation of Eden and tourism in a biblical context is uniquely problematic. One of the few details of the Genesis Garden of Eden story that practically everyone agrees upon is that humans are irrevocably barred from returning to the Garden. God placed cherubim with flaming swords at its entrance to keep the Tree of Life from humankind's grasp. The Eden myth serves many uses – as origin of human life, place of paradisical perfection or land cursed by original sin. But to visit an earthly Eden – even to locate one on a map – is impossibility or heresy.

Even for those who conceive of Christ as a 'Second Adam', belief in whom allows them to undo the fall of the first Adam, Christian salvation is rarely figured as a return to the biblical Garden of Eden, but more often as an ascension to a future, heavenly paradise, the 'new heaven and new earth' of Revelation. Nevertheless, there does exist a long history in Christianity of mapping the Garden of Eden on earth. (Although Judaism and Islam share the Eden story, both of those traditions tend to view the story more symbolically; neither faith has an active tradition of seeking Eden on earth.) That long history emerges out of Gen. 2.10-14, which comprise a geographical description of the Garden.

A river flows out of Eden to water the Garden, and from there it divides and becomes four branches. The name of the first is Pishon; it is the one that flows around the whole land of Havilah, where there is gold; and the gold of that land is good; bdellium and onyx stone are there. The name of the second river is Gihon; it is the one that flows around the whole land of Cush. The name of the third river is Tigris, which flows east of Assyria. And the fourth river is the Euphrates' (Gen. 2.10-14, NSRV).

The very concreteness of the names of rivers and lands in this passage has been enough to tempt theologians and geographers. Since the sixteenth century, when John Calvin's edition of the popular *Bishops' Bible* was the first to include a reference map of the Middle East, Protestant tradition has placed Eden in what is now Iraq, home of the Tigris and Euphrates Rivers.[1] In the nineteenth century, archaeologists began to correlate sections of Genesis, including Gen. 2.10-14, to the mythology of pre-Sumerian cultures, who lived in southern Mesopotamia around 3000 BCE. Sumerian descriptions of a magical land called Dilmun seemed to match the idea of Eden as a place of perfection and plenty, far removed from daily life. The vast, ancient marshes of southern Iraq are also thought to be an inspiration for Eden. One modern Iraqi town nearby has a shrine for a tree said to be the Tree of Knowledge, pictured in Alessandro Scafi's excellent *Mapping Paradise* (2006). But since this is 'primeval history' from before the invention of writing, the Garden of Eden has, according to biblical archaeologist Eric Cline (2007), fallen off the list of sites available for serious investigation.

There will never be a way to definitively prove whether the story's tellers thought of Eden as a real place or a mythical one, and if the former, no way to determine a specific location. Thus any claim that Eden can be found, or returned to, is a theological rather than an archaeological or historical claim. It is an unorthodox theological claim at that, because it must find a way to undo Adam and Eve's expulsion.

This chapter examines three such claims by idiosyncratic thinkers in the United States between 1885 and 1956. Methodist minister and Boston University founder William Fairfield Warren's *Paradise Found: The Cradle of the Human Race at the North Pole* in 1885; Ohio German pietist preacher Landon West's 1901 pamphlet contending that a Native American earthwork, the Serpent Mound, was the mark of Eden; and libertarian politician Elvy E. Callaway's mapping of Eden in the Panhandle of northern Florida, where he opened Garden of Eden Park in 1956. The sites described by the latter two are now state parks, and thus tourist destinations. Collectively, the three claims authorize a narrative of redemption that is progressively more accessible to sinners, tourists and travelers.

Four commonalities about these American Eden claims should be kept in mind at the outset. First, they require, as a precondition, the unraveling of the American Calvinist orthodoxy, in which original sin was foundational. Mid-century American intellectual historian Perry Miller famously described Ralph Waldo Emerson as 'a [Calvinist Jonathan] Edwards in whom the concept of original sin has evaporated' (1956: 185). The evaporation of original sin is a footnote for Miller, a small example of a discontinuity within the broad continuities he is trying to trace between seventeenth- and nineteenth-century thought, Edwards and Emerson. But when Eden-seeking picks up in the late nineteenth century, the lack of original sin becomes instead the very ground of analysis.

Secondly, Eden-seekers insist on basing their geographies in a particular reading of the same Bible verses: Gen. 2.10-14. Indeed, these verses form the limit of the genre of speculative Eden geographies. While seekers' interpretive strategies vary widely, this set of verses still acts as a set of constraints. The biblical constraint tends to separate Eden claims from utopian manifestos that were common in the nineteenth-century United States, and grounded less specifically in Scriptural exegesis. As a corrollary observation: for the most part, these Eden locations remained the domain of charismatic individuals, and did not draw masses of followers.

The obvious exception to that generalization is Joseph Smith, a highly charismatic individual with many followers. Smith declared that the Garden of Eden was in Independence, Missouri, though he did so on prophetic rather than interpretive authority. When his early Mormons were kicked out of Independence, he christened their next Missouri settlement 'Adam-ondi-Ahman', or the place where Adam and Eve settled; it is now a church-run pilgrimage site.[2]

Third, Eden-seekers read Gen. 2.10-14 in the context of newly urgent questions about human origins. The processes of reckoning and harmonizing sparked by the 1859 publication of Charles Darwin's *Origin of Species* intensified in the late decades of the nineteenth and early twentieth centuries. The Garden of Eden narrative comprises at least four essential narrative tropes – origins, perfection, sin and exile – and every reading of it attributes importance to these tropes in differing proportion. While these three seekers *see* all four aspects of the story, they can be said to be particularly focused on the first. For them, Eden geography has become important as a creation myth, an origin story for humanity or the starting point from which humanity emerged and spread.

Fourth and finally, these Eden claims, necessarily biblical and explicitly religious though their writers intended them to be, show a distinct lack of the contemporary American evangelical Christian preoccupation with biblical 'literalism' or 'inerrancy'. Such concerns were indeed fomenting in the late nineteenth-century 'Bible Wars', as a response to German 'higher criticism' and other perceived threats to the unity of Scripture. But the institutionalization of biblical literalism as a reading method didn't reach full saturation until 1909, when the publication of the Scofield Bible helped popularize Nelson Darby's 'dispensational premillennialism'. The Eden geography of young-earth creationists also incorporates 'flood geology', the pseudo-scientific biblical reading method codified by Whitcomb and Morris's 1961 *The Genesis Flood*. To wit: Eden had to have existed on Earth, because the Bible is inerrant. However, because of the destructive power of Noah's Flood, no one could ever presume to know where on earth it had been.

In the time period in question, however, Eden claims exemplify less of a theologically conservative effort to literalize Scripture than an Enlightenment inheritance of discovery, and a belief in human and historical progression. The combination of the 'evaporation' of original sin already underway, new urgency for questions of human origins, continued reliance on biblical authority, and absence of the institutionalization of biblical literalism made the late nineteenth and early twentieth centuries a particularly fertile interpretive environment, in which Eden geographies could thrive. (Elvy E. Callaway falls outside of this timeframe but still operates under the older dispensation, making him an isolated and contrarian figure in his own time, the 1950s.)

During this time, people continued to create new, ever more fanciful maps of Eden. They tended to acknowledge the Bible's mention of the Tigris and Euphrates, but to pay more attention to the other two rivers of paradise, the Pishon and the Gihon. Since those rivers have yet to be definitively located, they open up interpretive possibilities for anyone trying to locate Eden outside of Mesopotamia. For instance, British general Charles 'Chinese' Gordon tried to prove in 1886 that the island of Praslin, in the Seychelles, conformed to the biblical description. In 1914, Chinese Christian revolutionary Tse Tsan Tai published a hand-drawn map locating the four rivers in Outer Mongolia. Sir William Willcocks, British irrigation engineer, published a popular book called *From the Garden of Eden to the Crossing of the Jordan* in 1919 after his expedition to Iraq.

The Eden-seekers I have found in this period are determinedly idiosyncratic and difficult to generalize about. Most of them are male. An important exception would be famed early feminist Victoria Woodhull, whose 1875 lecture 'The Garden of Eden' (later published as Chapter 3 in *The Human Body the Temple of God*, London, 1890), dismisses all these male speculative geographies and instead locates the Garden inside the human body. Most of the Eden-seekers have some kind of footing in a Protestant tradition. Both William Fairfield Warren and Landon West were ordained clergy, Methodist and German Baptist, respectively. Elvy E.

Callaway's father had been a Baptist preacher, although he disavowed that particular church. But otherwise, they do not have much in common. That idiosyncracy is not accidental. Eden geographies are also cosmologies. By definition, each Eden-seeker believed themselves to be narrating the story of the creation and progress of all of human civilization, an enterprise that involved a lot of egotistical investment. If they did engage at all with the many other ideas about Eden's location that had long been circulated, they did so in order to dismiss them completely.

For this reason, the selection of three American Eden-seekers that I will focus on here is necessarily somewhat arbitrary, and the connections I draw between them not ones they would ever draw themselves. But the grouping does, I hope, provide a framework with which to make visible the Enlightenment inheritance of Eden-seeking in a late nineteenth- and early twentieth-century American context.

Arctic Eden

Today, William Fairfield Warren remains much better known as the first President of Boston University, and as a Methodist minister, than as an Eden-seeker. But in 1885, his book *Paradise Found* received wide public attention for its outlandish premise: the Garden of Eden, the 'cradle of the human race', had been at the North Pole. Warren dedicated his 500-page tome to the foremost German ethnographer, Max Muller (100), and proceeded to use the new method of 'comparative mythology' (x) to test his 'fresh hypothesis' (xiv).

Comparative mythology allowed Warren, for example, to read the word 'Euphrates' in Gen. 2.10-14 for its Greek root meaning 'deep', rather than as a proper name of a still-extant river in Mesopotamia. Along with the more customary 'Akkadian, Assyrian and Babylonian' traditions (xiv–xvii), Warren mined the mythology of Japan, China, Iran, Egypt and elsewhere, into his theory. Marshalling 'evidence' from these multiple traditions for each key element of the Eden story – the four rivers, the central tree (or trees), the profusion of life, the guardians at the gate – lent Warren's theory greater weight.

Then he simply correlated this comparative catalog of mythological evidence against the 'latest polar research', and arrived at the following synthesis. In prehistoric times, scientists had determined, polar regions would have been much warmer than they were in 1885. Explorers were finding fossils of woolly mammoths and large, sequoia-like trees. How much more of a leap was it to imagine the existence of primeval people?

Adam and Eve – or 'hyperborean Eocene man', as Warren called his first race – must have been 'of extraordinary stature and strength', like the flora and fauna (283). They lived in immortal peace and harmony, in a symmetrical world divided evenly into quadrants by the four rivers. One day, an unspecified disobedience on the part of hyperborean Eocene man caused God to flood this perfect world, pushing hyperborean Eocene man gradually farther and farther south, as the pole froze behind them. The farther they got from the Polar Eden, the smaller and more mortal they became.

The story of Eden was, for Warren, really the story of 'how the earth was peopled' (xxiv). It contained both an acknowledgement and an explicit religious critique of Darwinian thought, as perceived in the late nineteenth century. The acknowledgement lay in Warren's implicit reliance on scientific discourse for what was essentially a religious message. And the critique was one widely shared at the time. The idea of an 'ascent' of man from what Warren called, disapprovingly, 'sub-savage stupidity' (408) was incompatible with the Genesis depiction of God creating human Adam and Eve in 'His image'. So Warren figured his story, instead, as a 'descent' of man: from giant stature and Godly power to our degenerated current state.

The story of 'descent' was also an argument in an ongoing debate about whether or not the different 'races' of man had different geographical origins. As Warren put it, 'The problem of the original home of the human race is not [only] a question of Hebrew exegesis – it is a race-problem' (34). If different human 'races' originated from different places, their differences could be definitively essentialized. Warren was firmly on the side of the unified vision. For Warren, racial difference amounted to a difference in distance from some symmetrical, originary point.

This position was consistent with Warren's Methodist missionary impulse. To say that humanity shared a common origin point was to say that they were all equally in need of, and available to, Christian evangelizing. On the cusp of the discourse that Tomoko Masuzawa traces in *The Invention of World Religions* (2005), Warren would still proclaim Christianity was the 'world-religion', as indeed he did in his 1911 book, *The Religions of the World and the World-Religion*. For all Warren's apparent universalism, such as his drawing upon Greek, Roman, Sumerian, even Hindu folklore to back up his location of Eden at the North Pole, all of this evidence was being marshalled to endorse a progressively 'civilizing' pattern of humanity, evolving toward – or rather, back to – Enlightened Christianity.

Warren's argument about the peopling of the earth from a single Polar origin point also implicitly participated in a public debate over Polar exploration. During the years that he was developing his Eden theory, no one had successfully reached the North Pole, despite numerous expensive expeditions. Public enthusiasm for this great discovery dramatically waxed and waned. In 1884, the disastrous disappearance of the much-vaunted Greely Expedition to the Pole looked to have permanently soured public opinion against further exploration. Warren urged perseverance.

The Pole was not only the last prize in the long 'Age of Discovery' filling-in of geographical blank spots, it was also, Warren wrote in 1885, an important source of empirical knowledge. Polar exploration was 'almost certain to give us facts of inestimable value both to natural science and to archaeology' (491). The aurora borealis, and the electromagnetic fields around the earth, were additional sources of fascination. To monitor them, Warren proposed setting up an international research station at the cradle of the human race.

With his scientific-missionary impulse thus established, where does Warren's Polar Eden fall on our scale of accessibility? The question requires describing Warren's view of sin. As a Methodist, Warren subscribed to the Wesleyan theology of human perfectability, which had already gone a long way toward un-biting the apple of original sin, so to speak. (Some of Wesley's critics certainly thought it had gone too far.) In Warren's ornately detailed Polar Eden vision, he was uncharacteristically vague in his explanation of what the actual offending act that drove man out of the Garden was; far be it from him to reify the Augustinian attachment of sex and sin.

But Adam's fall was not as easy for Warren to overcome as Wesley's critics might have feared. Perfection may be possible, but the human sinner still must go through a lifelong struggle of justification and sanctification before reaching it. Warren's Polar Eden, likewise, was possible for humans to reach. But it was still important to Warren that reaching perfect Eden be *difficult*, and located in a place still untouched by humanity. We should be able to come back from sin, but not without a great effort.

Indeed, even the eventual arrival of American explorers at the North Pole was never conceived by Warren to represent a return to an *unspoiled* Eden. On the last pages of his book, before the numerous appendices, Warren writes: 'long-lost Eden is found, but its gates are barred against us. Now, as at the beginning of our exile, a sword turns every way to keep the Way of the Tree of Life'. In his conception, the Garden, abandoned, flooded and frozen, 'is Eden no longer' (432).

The intrepid explorer, Warren imagines, becomes more like a tragic pilgrim. 'Even could some new Columbus penetrate to the secret center of this Wonderland of the Ages, he could but hurriedly kneel amid a frozen desolation and, dumb with a nameless awe, let fall a few hot tears above the buried and desolated hearthstone of Humanity's earliest and loveliest home' (433). The only consolation for this new Columbus would be thoughts of the heavenly paradise available through Christ. That is to say, Warren's North Pole Eden was not really meant to be an earthly paradise, but as an object lesson in human history, one that would orient us toward a unified future, in terms of both science and salvation.

William Warren's existing platform as Boston University president, theologian and author meant that people did pay attention to his exhaustive, comparative-mythology account. *Paradise Found* was dutifully reviewed by *The New York Times* and the *Atlantic Monthly* in 1885. With every notice the book received, Warren enhanced his status as a go-to lecturer on the North Pole. By 1891, a *Boston Tribune* profile of Warren ('Men You Ought to Know!') claimed that 'thousands' believed in his North Pole theory.

Warren had intended his efforts to be the last, authoritative word on the location of Eden. However, perhaps because of its anticlimactic ending, *Paradise Found* provoked more questions than answers. Just as Warren had done with so many other sources and experts, new Eden-seekers, such as those discussed in a 1912 *Washington Post* article, still cited Warren as an authoritative backup for their Eden claims, claims that continued to pop up, undeterred, far into the twentieth century.

Ohio Eden

Reverend Edmond Landon West, a minister in the pietist German Baptist Brethren community of central Ohio, proposed his location for Eden in 1908. Like Warren, Reverend West was responding both to a particular modern condition, as well as to a very concrete local question. He wanted to know the provenance and purpose of the waist-high, serpent-shaped mound of earth that stretched more than 1,300 feet along the cliffs above the river in his hometown of Brush Creek. At one end of the mound, the serpent's tail was coiled, and at the other, its head, 120 feet long by itself, had jaws that looked to be about to swallow a round object.

According to West's 1908 pamphlet, 'Eden's Land and Garden with their Marks Yet to be Seen', the Serpent Mound was a depiction of the serpent of Genesis, tail writhing in agony as it consumed the poisonous fruit of sin. The earthwork was a 'mammoth pictured lesson' that the wages of sin are death (3). West denied the customary attribution of the earthwork to Native American builders, and insisted that it had been built by God Himself – or by the first people of Eden directed by Him – immediately after the expulsion, as a warning. Why else would the serpent be shown eating the forbidden fruit, when, as West noted, 'we should all know...that serpents don't eat fruit' (6).

Dutifully biblical, West located the requisite 'four rivers' of Genesis slightly upstream from the mound, where three tributaries joined to form the fourth, the Brush Creek. He knew that most people thought the events of Genesis happened in Asia, but he noted, accurately, that Moses (the traditional author of the Hebrew Bible) never wrote the word 'Asia' in Scripture. Reverend West calculated the time it would have taken for Noah to float by boat from Ohio to the, presumably biblical, Mount Ararat in Armenia, deeming it possible. (Though the Flood would have covered the Serpent Mound, when floodwaters receded, they washed away the earth that had concealed God's warning during the period between the Expulsion and the Flood, revealing the mark of Eden.)

West was very willing to incorporate available empirical evidence when it supported his biblical theory. When Harvard anthropologist F. W. Putnam's soil tests determined that the mound was much older than previously thought, West used this fact as evidence for his pre-Flood vintage of his Edenic tribe. In order to determine that the serpent mound represented the 'real' serpent in the Garden of Eden, and not the symbolic creature some were proposing, West went out and measured actual local snakes; he found their proportions corresponded to the snake depicted in the earthwork.

It was not coincidence, wrote West, that the geographical marker of the actual location of original sin was in what was now called 'Adams County', Ohio. This 'one mark of our history', had been 'so well made' and preserved 'to remind people of man's first sin... [It] is here yet to be seen by Bible readers and all who give it a visit.' The serpent lesson's job was to 'teach and give faith to all people' (11). And as such, the serpent needed to be seen.

The earthwork could only serve West's purposes if it were open and accessible, unlike Eden itself. One of the largest such mounds in the United States, Serpent Mound had been left in disrepair during West's childhood in the 1840s, but by the 1880s it had been discovered by Putnam, an anthropologist with Harvard's famed Peabody Museum, who drummed up interest among the upper class of Boston to raise funds for its restoration and preservation.

In 1900, Putnam turned over ownership of the site to the Ohio Historical Society. The Society's director, E. O. Randall, had heard West's theory, and in 1902 he quoted it in the organization's official bulletin, *Ohio Archaeological and Historical Publications*, with the caveat that West's idea 'was not exactly archaeology nor history though it contains something of each. It is, however, so unique and entertaining that we reproduce it as…food for the "higher critics", the Egyptologists archaeologists and the Biblical students of all classes'. Randall knew there were benefits to keeping West's entertaining theory in the public eye. With publicity, Randall could raise enough money to add a picnic pavilion to the Serpent Mound park, and, in 1908, an iron viewing tower, allowing visitors to see the whole length of the serpent at once. It was this last development, and the anticipation of numerous visitors that it represented, that may in turn have finally inspired Reverend West to write down his theory, in an 18-page pamphlet titled 'Eden's Land and Garden with Her Marks Still to Be Seen'. West wanted people to see those marks, like he had, and take them as a warning to sin no more.

West, like Warren, was not so specific about the nature of Adam and Eve's original sin, except to characterize 'that one sad event' as disobeying God. But he was very specific about the kinds of sins that the faithful, like himself, were still battling in 1908. Most of those evils could be categorized, for West, under the heading of 'division'. The first and most formative division for West was of course the American Civil War, which began when West was 19 years old. For him, war brought a particular quandary.

He belonged to the German Baptist Brethren, descendants of a German Pietist movement founded by the Schwarzenau Brethren in 1708. They were, along with the Mennonites and Quakers, one of three historic 'peace churches', which believed war to be categorically unacceptable in the eyes of God. However, many Brethren (also known as 'Dunkards') also believed that slavery was so grave a sin as to be worth fighting against. West devised a compromise: he was too sick with typhoid himself to go and fight, but he paid a substitute to do so for him, a relatively common practice.

West's country had divided and re-formed itself; so too had his family. He had three children with his first wife, and after her death five more with a second. His church had likewise weathered a threefold split, in 1881, into the conservative Old German Baptist Brethren, the liberal Brethren Church and the middle-of-the-road German Baptist Brethren. (The church would split again in 1926, and drop both the 'German' and 'Baptist' to become just the 'Brethren'.)

West had both reformist and traditionalist sensibilities. He advocated for the right of 'the colored', who had suffered at the hands of slavery, to be welcomed into the Brethren. He also advocated limiting communion and 'love feasts' to members of the church. He was against having a paid ministry; he was in favor, however, of Sunday School for children. So he stayed in the middle-ground Brethren, though ill health eventually made him less active than he had been in his fire-and-brimstone youth.

Even West's hometown – Brush Creek Township – had re-formed itself in his lifetime. In his childhood it had consisted of just a few farming families who belonged to four German Baptist churches; now there were, West wrote in his 1908 pamphlet, 'over 500 styles of religion' in the world, and fifteen houses of worship in Brush Creek Township alone, thanks to the new network of canals and state roads (8). Brush Creek Township had even been renamed; it was now called Peebles.

West may not have approved of all of these changes and divisions, but he did acknowledge that new conditions necessitated new religious initiatives. A younger West had written, in a tract against open communion, that 'The Bible is our main witness…we need all of it, and we need not a word more to insure to us the favor of Heaven's King' (1888: 68). Twenty years later in 1908, he confessed, speaking from experience, that 'the Bible cannot unite the people on any one thing' (8). Indeed, the Serpent Mound was 'the only mark of unity that our world, with its nations, races, and religions, now has' (8). As such, it was imperative that the unusual earthwork be seen by as many people as possible.

Importantly, West was not claiming that the pastoral Ohio hillsides of Adam's County were actually still Edenic; that would counteract his emphasis on humankind's sin. But he was nonetheless urging visitors to come and see the remarkable snake figure. If William Warren had envisioned the explorer who would eventually reach the North Pole as a lone Columbus 'letting fall a few hot tears' at the lost Edenic paradise, West saw a somewhat more cheerful vision: ordinary people arriving at the Serpent Mound by the dozens, with families and picnics. They could drive right up to the gates of the new preserve, climb the viewing tower and contemplate the mysterious landmark.

The tourists should come away realizing that all humanity was united in 'one line of descent that goes back to creation' and convinced of the imperative to 'st[and] united in loyalty to God's law' (8). The serpent was both a warning and an object of awe, an inducement to new faith. West apparently saw no conflict between the mark of Eden and a robust tourist trade. His tourists might be more like pilgrims, but there were certainly no cherubim with flaming swords to keep them away.

Instead, West often marveled at the mound's longevity, accessibility, and (to him) transparency. These qualities were assets: not unlike a gigantic, ancient Sunday School. Again disagreeing with his earlier stance on closing Dunkard communion, in 1908 West maintained that the serpent and its lessons should be open to everyone, which it soon would be as a state park. Today the Serpent Mound is managed by the nonprofit Arc of Appalachia, which in its 2015 IRS Form 990 reported 33,521 visitors. West's picture and pamphlet are still part of the interpretive materials at the site.

Warren's Eden was, for all its scientific and missionary enticements, still essentially a frozen and inaccessible wilderness, abandoned by sinful humans. West had to frame the Serpent Mound as a mark of sinfulness, rather than a mark of paradise, in order to justify its public accessibility. Florida lawyer Elvy E. Callaway, in contrast, eliminated sin from the Eden equation completely, and in doing so most definitively opened up his Eden to American visitors.

Florida Eden

Callaway's rejection of original sin was emphatic and complete. As he wrote in his 1966 book *In the Beginning: Creation, Evolution, The Garden of Eden, and Noah's Ark*, 'I have often heard it preached that we are all born in sin – natural crooks. It is a slander of God to teach any such doctrine' (40). What, then, had happened in the Garden of Eden? Callaway, shrewdly incorporating the fact that there are two creation stories in the book of Genesis, argued that God first created humans without souls or the power of choice. (Those people eventually died out, but not before providing an answer to the age-old question of how Cain was able to find a wife so soon after being expelled from the Garden.)

But in the second creation, He thought better of this, creating Adam and Eve, and endowing them equally with what Callaway calls 'the right of liberty' (39), the power to make their own decisions. 'Mother Eve' knew that she had to eat from the Tree of Knowledge in order to activate this liberty, and by doing so she 'deliberately chose to come under the law that ultimately meant...the death of their bodies, in order that they might have a role in the production of something better' (38).

Immortality meant stagnation, and Callaway praises Mother Eve for her 'marvelous choice' (39) to embrace progress instead. The real threat in Callaway's story is 'not a serpent, but a Communist or a welfare-statist' (37), who would take away that right of liberty. Callaway had fused a sort of 1930s' progressivism that favored knowledge and education (and, not incidentally, women's rights), with a classic libertarianism that was suspicious of government involvement in such progress.

Since there was officially no sin, there was also no exile in Callaway's story. Adam simply decided to leave the Garden one day, and Eve followed him. The cherubim turning their flaming swords were thus not meant to keep people away from the Garden but to protect Eden's sacred trees of liberty and knowledge from disruption by communists or welfare-statists. Everyone else, however, was welcome to come and visit, for the low-ticket price of one dollar and ten cents.

In 1956, according to *Chicago Daily Tribune* travel columnist Thomas Morrow, Callaway set up a 'Garden of Eden Park' along the two-lane Highway 12 near Bristol, Florida, along the Apalachicola River in the Florida Panhandle. Once visitors paid their admission at a roadside ticket kiosk and toured the small museum, they could head out on a 3.75-mile hiking trail to observe the natural wonders of the Apalachicola.

Edenic dreams are of course nothing new in Florida, where Ponce de Leon once located the Fountain of Youth, and where, Bruce Boehrer (2006) reminds us, the Civil War-era governor John Milton was believed for generations to be a direct relative of the author of *Paradise Lost*. But Callaway's Eden was on the wrong side of Florida – the North Florida Panhandle was rarely the site of such visions as the beautiful southern beaches of Miami and Orlando. An unauthored 1973 editorial in the *St. Petersburg Times* tacitly acknowledged this imbalance when it allowed that Callaway's Garden of Eden Park, though bizarre, would be 'a tourist boon to the state's least populous county' (177). There are no tourist numbers for this period, but the Park did receive some attention.

Just after the park opened, columnist Thomas Morrow (1956) interviewed an elderly couple, Mr. and Mrs. F. W. Wentworth from Valdosta, Georgia, a two-hour drive away. Travel writers from Gloria Jahoda (1967) to Rory McLean (2000) describe the park, or what was left of it, as the embodiment of one charismatic individual, although an object of a certain amount of local pride, rather than a large-scale destination. (Callaway died in 1981, and although his 'Garden of Eden Park' is still remembered in travelers' accounts, it is no longer officially in existence. However, the land it stood on has been preserved via its incorporation into Florida's Torreya State Park.)

The 'wrong' location and limited success with tourists was exactly the sort of challenge Callaway was up for. He was constitutionally contrarian in every way – personally, religiously and politically. Callaway knew his Garden of Eden Park needed constant promotion, and he seized every opportunity to do so. For example, according to Jahoda, after the conservative Barry Goldwater lost his campaign for President to Richard Nixon in 1964, Callaway publicly offered his idol a home in the Garden of Eden as a consolation prize (1967: 70).

The strange Floridian theology visible in the Garden of Eden Park can be seen in nascent form in Callaway's 1934 book *The Other Side of the South*. A proud Southerner, he 'blessed the North' for preserving the Union during the Civil War, and dedicated his book to those Southerners who had not supported the Confederacy. He proudly told the story of being ousted from his tiny hometown Baptist church in Weogufka, Alabama, for dancing with a local schoolteacher, and claimed support for the skeptic Clarence Darrow in the 1925 Scopes Trial (61).

Politically, Callaway was a Republican, and an active one. He had served as a Republican delegate to the Convention that elected Herbert Hoover President, as he reminded the editors of the *Washington Post* in a 1928 letter, and served for years as Florida's Republican Party chairman. It is difficult to overstate how out-of-place a Republican would have been in the [American] South during the Great Depression of the 1920s and 30s, the years of the 'Dixie Democrat' party machine. Callaway wrote, not inaccurately but with considerable hyperbole, that any Floridian who questions a Democrat 'commits an unpardonable sin and is looked upon as a traitor to all that is sacred and holy'.

He hated President Franklin Delano Roosevelt's government-heavy New Deal, which began in 1932, and he denounced those Democratic policies in his 1936 campaign for governor of Florida. (It was, ironically, the New Deal-era Civilian Conservation Corps that first began preserving the Apalachicola land now encompassed in the Torreya State Park, which was already in place by the time Callaway wrote *In the Beginning*.)

Callaway lost, dramatically. And yet, the kinds of reform and development projects Callaway argued for are not so different from those that Roosevelt had in mind: government assistance of business in agriculture, improved public schooling, especially for 'the Negro', and a general overturning of stereotypes of the American south as backward, uneducated, and stuck in the Confederate past. If Southerners could just achieve those goals, wrote Callaway in *The Other Side of the South*, Florida would become 'the Eldorado of this continent' (1934: 169). The progression Callaway made from his 1934 'Eldorado' to his 1956 'Garden of Eden' is a strange one for an avowed religious skeptic.

Religious skeptic though he was, Callaway still framed his Garden of Eden theory like every other Eden-seeker before him, with the four rivers. Callaway did not have to look far. Just north of the park, along the Florida–Georgia border, the Jim Woodruff Dam marked the confluence of a four-headed river system, and 'there is no other river system in the world that meets the Bible description' (Callaway 1966; frontispiece). The Tigris was the Chattahoochee, Spring Creek the Euphrates, Fish Pond Creek the Pishon and Flint River the Gihon.

Like Warren and West, Callaway looked toward scientific authority to legitimate his theory. He quoted a local geologist to the effect that the landmass at the southern edge of the Appalachian mountain chain was one of the oldest in the world. Like many Eden-seekers, Callaway was also preoccupied by its bookend event in Genesis, Noah's Flood. He consulted specialists to confirm the biological uniqueness of the *Torreya taxifolia*, an endangered tree species he had first encountered in surveying his Bristol land in 1952. In 1966, Callaway argued this tree was equivalent to the 'gopher wood' the Bible says Noah used to build his Ark. He commissioned an Ark replica made out of Torreya wood, which is what he displayed at the small museum. To prove the Flood connection, he did invoke stranger lines of evidence than geology and biology, including extensive etymological and numerological analysis, based on ideas that he attributed to one Dr. Brown Landone, who, he claims, taught him about the secret priesthood of Melchizedek, and suggested he look north for answers – which he found in north Florida (75).

Callaway's insistence on framing what were essentially progressive ideals for the improvement of the South in biblical terms is perhaps understandable if viewed in the same spirit as his other contrarian turns. Although he knew his Republican gubernatorial campaign was futile in a Democratic South, he refused to let the 'machine' run unopposed. Although Callaway felt the Southern church had abandoned true religion for backwardness and superstition that was impeding progress, he was unwilling to cede the ground of religious language entirely – even if to do so he had to mix that old-time religion with modern science, local boosterism, libertarian politics and mystical numerology.

The balance Callaway sought between 'religion and reason' was one that the three Eden-seekers here share (1934: 51). Speculative geographies of Eden are highly syncretic endeavors. Ironically, although these thinkers are drawing geographical lines, they are entirely unconcerned with disciplinary boundaries rules of evidence. Because the motivations behind these speculations are individually urgent to each Eden-seeker promoting them, evidence of any kind available must be used – whether biblical, mystical or empirical.

Conclusion

Perhaps the act of harmonizing the mystical with the empirical was itself effective as a performative act; an exercise to show it could be done, despite all apparent conflict between the two in a modernizing and

Enlightenment-influenced America. Warren embraced the new science of Polar exploration to make an old missionary plea. West saw an opportunity to make the old, 'historical' story of Eden present for a modern tourist audience that still needed to hear it. West and Elvy Callaway, although they were at opposite ends of the spectrum of religiosity, and at opposite ends of the Appalachian mountain range, still both felt the need to calculate the rate of speed at which Noah could have floated from the Americas to the Middle East (Callaway's rate was faster). All three men saw themselves as performing a necessary integration of religion and reason, one that could provide a service to confused modern Americans.

Although these are modern cosmologies, a form of poetic worldmaking or folk theology, their efficacy cannot be measured in the number of books sold or adherents converted. As a genre, nineteenth-century American Eden theories can appear quaint, or even desperate, to twenty-first-century readers. But if the painstakingly syncretic logic by which Warren, West and Callaway justified their speculative geographies has long since evaporated like Miller's original sin, we are left still with their actual geographies.

Warren's Polar Eden has of course now been thoroughly explored; the aurora borealis and magnetic field thoroughly researched. Both West and Callaway were building speculative geographies around landscapes in their own backyards, so to speak, and landscapes which were already in the process of being preserved for posterity when their theories were published. All three men had every reason to anticipate future visitors to their respective Edens. Warren's Eden theory has not been memorialized at the Pole, but his comparative mythology method, much as he intended it to be the definitive last word on the problem of Eden, in fact extended the lifespan of the Eden-seeking genre by several decades at least, with authors with very different final conclusions citing Warren as authoritative scholarship.

West and Callaway are both remembered at the physical sites that they so fervently advocated for. West is pictured and described in the interpretive materials at Serpent Mound State Park in Ohio; its caretakers have continued E. O. Randall's practice of incorporating every possible reason to visit the site into their message. Callaway is not officially acknowledged in the Torreya State Park system that his Garden of Eden Park was once adjacent to, but has remained a folk legend, continuously noted by travel writers and acknowledged informally by park staff, especially in reference to explaining the endangered Torreya taxifolia tree, for which the park is named. Here too, Callaway's 'gopher wood' mythology has been marshalled alongside the actual biological uniqueness of the tree as both a curiosity and as an additional reason to conserve the area.

In these contexts, most of the key functions of the Garden of Eden are lost – the origin story, the warning of sin and the tale of exile. What is preserved, however, is still a certain amount of the other element of the story: perfection. Visitors, then and now, likely do not know about the Eden theories before they arrive at the Serpent Mound or Torreya State Park. But once they arrive, the vestiges of West and Callaway's quixotic biblical interpretations still function as part of the overall appeal to visitors.

These tourists are not making pilgrimages to holy sites – as, for example, the contemporary Mormon tourists that visit Adam-ondi-Ahman are. They do not have to navigate past cherubim with flaming swords. That does not mean, however, that there is no added value to their touristic experience of visiting a place that someone once, long ago, thought they could prove was the Garden of Eden. The very disjunction between the idea of Eden and the reality of Ohio or Florida invites curiosity. How, we wonder, did these historical characters possibly get from one to the other? There may even be some political value, within contemporary American culture, to preserving these vestiges of a richer set of biblical interpretations, keeping alive some of the interpretive fertility of the late nineteenth and early twentieth centuries. Visiting these Edens does not augur our future, but invites us to appreciate the theological creativity of the not-so-distant past.

Notes

1. For more on the pre-modern Eden search, see Jean Delumeau's *A History of Paradise* (1995).
2. While Mormons are outside the scope of this chapter, they are discussed extensively in my 2011 book about Eden-seeking, Paradise Lust.

Bibliography

Arc of Appalachia IRS Form 990 (2015), http://arcofappalachia.org/wp-content/uploads/2015/04/2015-IRS-990.pdf

Boehrer, B. (2006), 'Milton in America: The Case of Jackson County, Florida', *Textual Practice*, 20 (1): 1–24.

Callaway, E. E. (1928), 'Letters to the Editor: Comment on the Pepper Victory', *The Washington Post*, May 11, ×8.

Callaway, E. E. (1934), *The Other Side of the South*, Chicago: Daniel Ryerson.

Callaway, E. E. (1966), *In the Beginning: Creation, Evolution, The Garden of Eden, and Noah's Ark*, New York: Carlton Press/Reflection Books.

Cline, E. (2007), *From Eden to Exile: Unraveling Mysteries of the Bible*, Washington, D.C.: National Geographic Society.

Delumeau, J. (1995), *A History of Paradise: The Garden of Eden in Myth and Tradition*, New York: Continuum.
Editorial (1973), 'Visit the Garden of Eden', *St. Petersburg* [FL] *Times*, 29 July, 177.
Gordon, H. W. (1886), *Events in the Life of Charles George Gordon from Its Beginning to Its End*, London: Kegan Paul, Trench.
Jahoda, G. (1967), *The Other Florida*, New York: Charles Scribner's Sons.
Masuzawa, T. (2005), *The Invention of World Religions, Or, How European Universalism Was Preserved in the Language of Pluralism*, Chicago: University of Chicago Press.
MacLean, R. (2000), *Next Exit: Magic Kingdom: Florida Accidentally*, New York: Harper Collins.
'Men You Ought to Know!' (1891), *Boston Daily Globe*, 27 December, 9.
Miller, P. (1956), *Errand into the Wilderness*, Cambridge: Belknap Press.
'Modern Discovery is Dispelling the Weird Delusions About the World' (1912), *Washington Post*, 24 March.
Morrow, T. (1956), 'Want to Get Into Garden of Eden for Buck, 10 Cents?', *Chicago Daily Tribune*, 14 February, A2.
Randall, E. O. (1902), 'Ohio, the Site of the Garden of Eden… The Theory of Rev. Landon West', in *Ohio Archaeological and Historical Publications, Volume X*, 225, Columbus, OH: Published for the Society by Fred J. Heer.
'Review of *Paradise Found*' (1885), *The Atlantic Monthly*, July, 126–32.
Scafi, A. (2006), *Mapping Paradise: The History of Heaven on Earth*, Chicago: University of Chicago Press.
Tai, T. T. (1914), *The Creation: The Real Situation of Eden, and the Origin of the Chinese*, Hong Kong: Kelly & Walsh.
Warren, W. F. (1885), *Paradise Found: The Cradle of the Human Race at the North Pole: A Study of the Prehistoric World*, Boston: Houghton & Mifflin.
Warren, W. F. (1911), *The Religions of the World and the World-Religion: An Outline for Personal and Class Use*, New York: Eaton & Mains.
West, E. L. (1888), *Close Communion; or, Plea for the Dunkard People, in 2 Parts*, Dayton, OH: G. B. Brethren Book & Tract Work.
West, E. L. (1908), 'Eden's Land and Garden with their Marks Yet to be Seen'. Pleasant Hill, Miami Country, Ohio, 1908; American Theological Library Association [ATLA] Historical Monographs Collection: Series 2 [1894 to 1923]; Fiche 1990–5072). EBSCO*host* (accessed 30 March 2018).
'Where is Eden?' (1885), *New York Times*, 5 April, 5.
Whitcomb, J. C. Jr. and H. M. Morris (1961), *The Genesis Flood: The Biblical Record and its Scientific Implications*, Philadelphia: Presbyterian & Reformed.
Wilensky-Lanford, B. (2011), *Paradise Lust: Searching for the Garden of Eden*, New York: Grove Press.
Willcocks, W. (1919, 1920), *From the Garden of Eden to the Crossing of the Jordan*, 2nd ed., London: E. & F. N. Spon; New York: Spon & Chamberlain.
Woodhull, V. (1890), 'The Garden of Eden', in *The Human Body: The Temple of God*, London: 17 Hyde Park Gate, S.W.

Part II

Pilgrimage and Religious Tourism

Chapter 4

The Latter-Day Saints, the Bible and Tourism

Daniel H. Olsen and George A. Pierce

Introduction

Like adherents in many religious faiths, members of The Church of Jesus Christ of Latter-Day Saints (hereafter the Latter-Day Saint Church) hold certain places to be holy or set apart from other, more mundane, spaces. Like other Christian faiths, the Latter-Day Saint Church holds the Holy Land and places that mark the life and journeys of Jesus Christ to be sacred places (Hudman and Jackson 1992; Olsen 2006). As well, since the early twentieth century, Church leaders have systematically marked, purchased and maintained historical sites where important events related to the founding of the Church during the previous century took place (Erekson 2005; Olsen 2013). This was done both to preserve 'salvation history' (Davies 2000) – to witness and remember 'the living God-who-acts-in-history' (Tobler and Ellsworth 1992: 596) and his divine intervention in establishing the Church – and to construct or shape a standardized and collective Mormon religious and cultural identity through sacralizing these 'places of memory' (Nora 1989; see Madsen 2008; Laga 2010; Olsen 2013). While there is no formal 'theology of tourism' espoused by leaders of the Latter-Day Saint Church, every year thousands of Church members, generally with their families, engage in 'faith tourism', or 'non-pilgrimage multifaith style[s] of tourism to religious sites' (Stausberg 2011: 156), and visit these sacralized sites to build their own faith and transfer that faith to future generations (Lankford, Dieser and Walker 2006; Olsen 2006, 2013, 2016).

While there is growing literature related to the travel patterns of Latter-Day Saints and the management of Church religious heritage sites in the United States (e.g., Hudman and Jackson 1992; Madsen 2008; Olsen 2006, 2012a, 2012b, 2013, 2016, 2019; Brayley 2010; Schott 2010; Howlett 2014; Olsen and Timothy 2018), no research has focused on examining the underlying biblical themes that run through the narrative production and management of Latter-Day Saint religious heritage sites. To investigate the importance of the Bible and its theological precepts in the construction of Latter-Day Saint historical sites and their use in identity management, the first section of this chapter discusses the importance of the Bible in the historical and theological development of the Latter-Day Saint Church (see Barlow 2013) and the subsequent interest by Church leaders in the Holy Land. The rest of the chapter focuses on showing how biblical themes such as temples, prophets, scriptures, restoration and priesthood are interwoven into the historical events of particular Church religious heritage sites and the subsequent narratives surrounding modern interpretation of these events.

Latter-Day Saints and the Bible

From the beginnings of the Mormon movement, Joseph Smith took a literalist interpretation of the Bible, and eagerly sought to identify his religious movement with the fulfillment of biblical prophecy and a millenarian belief in being the chosen people of God, or an extension of biblical Israel, living in the final days of history (Underwood 2005: 44; Yorgason 2010). Joseph's familiarity with the Bible and other religious literature informed his thinking and provided background to the main tenets of what Latter-Day Saints term the 'Restoration'. This 'Restoration' included the belief that God continued to interact with humanity in biblical fashion, had restored the priesthood offices of prophet, apostles and elders, and that the gathering of the House of Israel was imminent.[1]

For Latter-Day Saints, the 'Restoration' was directly connected to Joseph's inquiry during the Second Great Awakening regarding which Christian denomination he should join (Wood 1980; Bushman 1984).[2] An appeal to the Bible for guidance and a subsequent reading of Jas 1.5, which encouraged readers who 'lack[ed] wisdom' to 'ask of God', led to Joseph Smith to pray in the woods near his home, where he received a hierophany or 'First Vision', wherein he was visited by God the Father and Jesus Christ and told to re-establish or restore Christ's primitive church.

Additionally, the Book of Mormon, Mormonism's foundational text, incorporates large portions of the books of Isaiah and Malachi and references several biblical prophecies and narratives. The language and cadence of the Book of Mormon was rendered in a King James-like style to foster a familiarity between the Book of Mormon text, the KJV Bible and religious discourse encountered by English-speaking Christians of the early nineteenth century (Belnap 2011). As well, the subtitle of the Book of Mormon, 'Another Testament of Jesus Christ', highlights the Latter-Day Saints view of the Book of Mormon as being a close companion to the Bible in witnessing to the reality and divinity of Jesus Christ. Indeed, Latter-Day Saints see the publication of the Book of Mormon as the fulfillment of biblical prophecy, where the Stick of Judah (the Bible) and the Stick of Joseph (the Book of Mormon) would join together (Ezek. 37.16-17), with its story of Israelites who traveled to the American continent akin to the branches of Joseph who 'run over the wall' in Gen. 49.22.

The connection between the Bible, biblical Israel and Joseph Smith's movement through the fulfillment of prophecy is evident in the early days of the church, as evinced in the visitation of the angel Moroni to Smith on the evening of 21 September 1823, the account of which was later canonized as Latter-Day Saint scripture.[3] During the night, Moroni appeared to Smith and informed him of his prophetic role and the existence of gold plates, the source of the Book of Mormon, buried in a nearby hill. Smith also related that Moroni quoted portions of Malachi 3–4 with slight emendations, Isaiah 11, Acts 3.22-23, Joel 2.28-32 and many other passages of biblical scripture (Joseph Smith, *History* 1:30-41). Subjects such as Elijah preceding the 'day of the Lord', the gathering and restoration of the House of Israel, Christ as the Moses-like prophet and the pouring out of the Spirit of the Lord all provided the scriptural basis for Latter-Day Saint millenarianism and constructed ties between the Saints, the House of Israel and ultimately the Holy Land as well.

Joseph Smith and his followers clearly saw themselves as living in the 'the dispensation of the fulness of times' (Eph. 1.10) that would immediately precede a millennial reign of Christ. In 1831, Joseph made a connection between the angel preaching the everlasting gospel in the end time and the ministry and mission of the Latter-Day Saint Church (Rev. 14.6-7; Doctrine and Covenants 133.36-40). In a letter dated 6 September 1842, later canonized as Section 128 of the Doctrine and Covenants, Joseph wrote about the founding of the Church and restoration of priesthood authority, and employed the phrase the 'dispensation of the fulness of times', and later referenced the 'fulness of Times' in connection with the vision of the 144,000 seen by John in Rev. 14.1-5 (Hedges, Smith and Anderson 2011).

Also, as part of the 'restitution of all things' (Acts 3.21) that must come to pass prior to Christ's Second Coming, Joseph Smith restored many divine–human covenants and rituals which could only be made and performed within a temple setting. These temple covenants, including the rituals of baptisms for the dead, the endowment,[4] and marriage sealings for time and eternity, were a part of the restoration of the 'fulness [*sic*] of the priesthood' (Doctrine and Covenants 124.26-28) and invite comparisons between biblical priesthood ritual and the Jerusalem temples and the Latter-Day Saint priesthood, ordinances and temples.[5]

Additionally, Smith felt deeply about the prevalent early nineteenth-century view of the United States as being a divinely appointed country (Stephanson 1995: 5; see also Doctrine and Covenants 101.76-80). This view, combined with the belief that Smith's movement was fulfilling biblical prophecy and the inheritor of Israel's covenant blessings, led Smith to tie certain biblical events to specific locations in the United States. For example, while visiting Missouri, Smith identified a location near Spring Hill, Daviess County, Missouri, as the place where Adam gathered all his righteous posterity from Seth to Methuselah and blessed them. Smith also declared that this place, which he called Adam-ondi-Ahman, was the place where 'Adam shall come to visit his people, or the Ancient of Days shall sit, as spoken of by Daniel the prophet' (Doctrine and Covenants 116.1; see Dan. 7.13-14, 22). Adam-ondi-Ahman was also where Christ would one day return, partake of the sacrament with the Patriarchs and other biblical figures and judge the righteous (Doctrine and Covenants 27.5-14; 107.53-57). Smith also identified Independence, Jackson County, Missouri as the location of the Garden of Eden[6] and the place where 'Zion', or the 'New Jerusalem' (Rev. 21.2, 10), would be built (Doctrine and Covenants 45.64-67; 57.1-3; 84.2-3).[7] With these declarations, Smith wove together these Adamic events with the American continent, providing a sense of biblical heritage and apocalyptic anticipation for early Latter-Day Saints settling in Missouri.[8]

While the connection between Joseph Smith's religious movement, events depicted in Isaiah and Revelation, Mount Zion and the United States, are clearly expressed through Smith's sermons and writings, ties to the Holy Land were first forged by Orson Hyde, who served as an apostle in the church, during a proselytizing mission in 1841. During his mission, Hyde was 'moved upon by the Spirit of the Lord to visit [Palestine], and stated that he intended to go to New York, London,

Amsterdam, Constantinople, and the Holy Land' (Smith and Thompson 1840: 92). While in Jerusalem, Hyde noted that many Jews were moving there from Europe and saw this in-migration as evidence that 'the great wheel is unquestionably in motion, and the word of the Almighty has declared that it shall roll' (Hyde 1842: 741) regarding the gathering of Israel. To this end, on 24 October 1841, Hyde ascended the Mount of Olives in Jerusalem, built a small stone monument and offered a prayer in which he dedicated the Holy Land as a place for the gathering of Israel (Hyde 1842; Van Dyke and Berrett 2008). Orson Hyde's mission established a link between Latter-Day Saints, Jerusalem and the Holy Land that exists through the present day by means of the Orson Hyde Memorial Garden on the Mount of Olives dedicated in 1979, and the Brigham Young University (BYU) Jerusalem Center for Near Eastern Studies on Mount Scopus opened in 1989 (Olsen and Guelke 2004).

Latter-Day Saints also maintain strong links through traveling to the Holy Land for touristic, religious and educational reasons. For example, Brigham Young University, owned by the Latter-Day Saint Church, has run a study abroad program since 1968 to the Holy Land, which continues through students attending classes at the BYU Jerusalem Center (Galbraith, Ogden and Skinner 1996; Olsen and Guelke 2004). Latter-Day Saints also study the Old and New Testaments as a part of their Sunday School curriculum, which inspires some Church members to engage in informal pilgrimage-like travel to the Holy Land. Because of this growing interest in travel to the Holy Land, as well as places related to the New Testament outside of the region, a small number of tour agencies, mostly located in Utah, have begun to focus on a specifically Latter-Day Saint clientele. These agencies offer tours to Christian sites in the Holy Land, including the Valley of Elah – where as a part of the experiential learning process many Latter-Day Saint tourists re-enact the David vs. Goliath battle (Bajc 2007) – and provide guides that are sympathetic to the Latter-Day Saint worldview in their interpretation of biblical sites (Olsen 2006).

Biblical Themes at Latter-Day Saint Religious Heritage Sites

In recent years, Christian groups in the United States have attempted to Christianize the American cultural landscape. Christian-themed messaging on highway billboards, religious bumper stickers, wayside shrines and crosses and religious messages painted on the sides of semi-trailers pervade many parts of the United States (Beal 2005; MacDowell 1982; Olsen 2015). As well, some religious groups use the North American religious tourism market to rematerialize the Bible, and do so in a way that combines popular culture, recreation and commodification within a fashionable and accepted religious and biblical narrative and landscape (Bielo 2016). Examples of this range from 'big things', such as the replica of Noah's Ark in Maryland and the world's largest large Ten Commandments monument, to religious theme parks, such as the Fields of the Wood religious theme park North Carolina and the Holy Land Experience theme park in Florida, to creationist museums, such as the Creation Museum and Ark Encounter in Kentucky and the Museum of Creation and Earth History in California (Lukens-Bull and Fafard 2007; Olsen 2015; Bielo 2017, 2018).

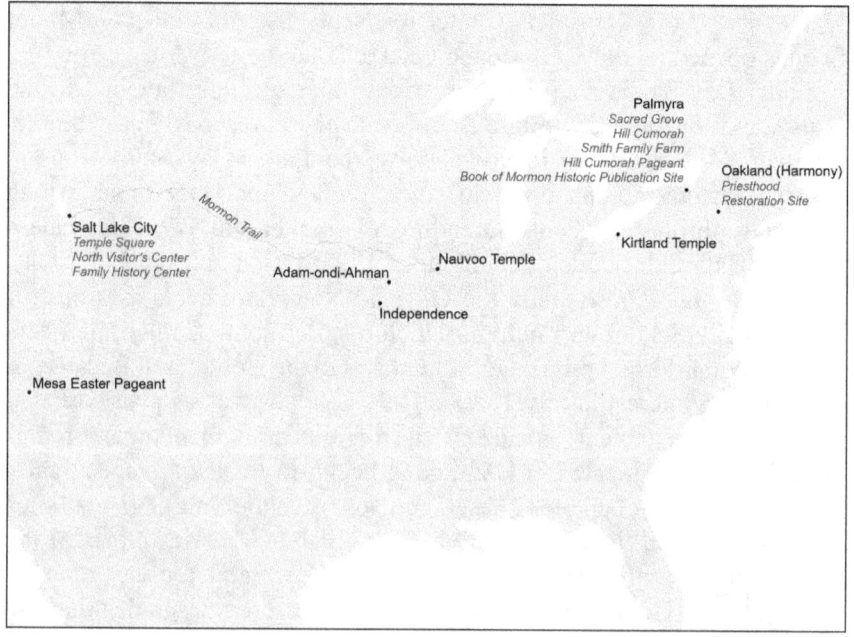

Figure 4.1. A map showing the major Latter-Day Saint religious heritage sites and routes in the United States. Source: Think Spatial, Brigham Young University.

For the Latter-Day Saint Church, rather than attempting to rematerialize the Bible and link faith foundations to events and people of the Bible, focus is placed on marking and commemorating places where significant historical events of the restoration, organization and development of the early Church in the 1800s took place. As noted above, these sites, which range geographically from Vermont and New York to

Ohio, Missouri, Illinois and Utah (see Fig. 4.1), are visited by thousands of Latter-Day Saints a year, and are used by Church leaders to remember and witness the role of God's hand in the restoration of the Church; to construct a broader Latter-Day Saint identity among Church members, especially for those living outside of the United States; to help construct a positive image of the Church to non-Latter-Day Saint visitors; and to further proselytization goals (Olsen 2013: 230–5; 2016). This is done not only by sharing the story of and witnessing to the truth-claims of the Latter-Day Saint Church at these sites, but also through highlighting the biblical foundations of the restoration of the Church as outlined by Joseph Smith and other Church leaders.

New York

As noted above, the foundational events of the Restoration occurred outside of Palmyra, New York, where Joseph Smith lived during his teenage years. The most important of these events took place in the Sacred Grove, a forested area close to Smith's home, where Joseph Smith had his 'First Vision'. While the exact location of this foundational event is not known, the forest in the approximate area where Smith prayed is owned by the Church and is considered 'holy ground', much like Mount Horeb where God visited Moses (Exod. 3.5) (Mitchell 2002). It is expected that visitors will be reverent when visiting the woods, as many Latter-Day Saints use this space to meditate and reflect on their religious testimony and identity. However, even with the stated sanctity of the site noted on a Church-sponsored sign, it is common for Latter-Day Saint visitors take a fallen leaf or a rock as a souvenir of their experiences at the grove (Anderson 1980).

The Hill Cumorah is located near the Smith Family Farm where he found the gold plates from which he translated the Book of Mormon. Smith stated that he was able to translate the plates through the use of special interpreter stones that were included with the plates, stones which were functionally similar to the Urim and Thummim, or the oracular means of revelation entrusted to the high priest as noted in Exod. 28.30 and Lev. 8.8. Egbert B. Grandin, who owned a bookshop and a printing press in Palmyra, printed the Book of Mormon manuscript. This was also the bookstore from which Smith purchased the Bible he used to engage in a re-translation of the Bible.[9] In 1978 the Church bought the building where the press was located and converted it into the Book of Mormon Historic Publication Site, which serves as a visitors' center to explain the history and process behind the publication of the Book of Mormon. Church leaders have also purchased the Smith Family Farm and have reconstructed the original home where the angel Moroni visited Smith and impressed on him the biblical ties to Smith's role as a prophet in the last days.

Pennsylvania

Joseph Smith moved from Palmyra to Harmony (now Oakland), Pennsylvania where he translated most of the Book of Mormon during a stay with his parents-in-law, the Hales. During the translation process, Smith, along with Oliver Cowdery who was serving as his scribe, had questions regarding baptism. On 15 May 1829, while praying in the woods about this topic, a resurrected John the Baptist appeared and bestowed upon them the Aaronic Priesthood and the keys to administer in the ordinance of baptism. Smith and Cowdery then went to the nearby Susquehanna River and baptized each other. Later that year, Peter, James and John, Christ's original apostles, appeared to Smith and Cowdery and gave them the keys to the Melchizedek Priesthood.[10] As such, this location is important for the restoration of priesthood keys and power in the Church.

Church leaders purchased the land where these events took place between 1947 and 1959, but it was not until 2015 that the Church fully developed the location as a religious historic site (Lloyd 2015; see Thayne 2007; Staker 2011). The Priesthood Restoration Site is 90 acres in size, and includes a visitors' center where a film providing an overview of this priesthood restoration event is shown, and Latter-Day Saint missionaries give tours of surrounding reconstructed historical buildings, including the Hale Farm and Smith Home where Smith translated most of the Book of Mormon. There are also two sculptures depicting Smith and Cowdery receiving the Aaronic and Melchizedek priesthoods from John the Baptist and Peter, James and John respectively.

Ohio

Because of religious persecution, Smith and his followers moved from New York to Kirtland, Ohio. While in Kirtland, Smith received a revelation to build a temple in preparation for the restoration of additional teachings and ordinances or religious rituals so the Saints could be endowed 'with power from on high' (Doctrine and Covenants 95.8). During the dedication of the Kirtland Temple, in March 1836 and in the days that followed, a number of Church leaders and members experienced spiritual manifestations. One such event was described by witnesses as a Day of Pentecost-like event (see Acts 2.1-4). On 27 March, the day of the temple dedication, one attendee wrote that 'The Spirit was poured out – I saw the glory of God, like a great cloud, come down and rest upon the house and fill the same like a mighty rushing wind. I also saw cloven tongues, like as of fire rest upon many, (for there were 316 present,) while they spake with other tongues and prophesied' (Arrington 1972: 426). A few days later, on 3 April, another miraculous event was recorded. During a meeting in the temple, Moses appeared and bestowed upon Smith 'the keys of the gathering of Israel from the four parts of the earth, and the leading of the ten tribes from the land of the north'. Then Elias appeared and 'committed the dispensation of the gospel of Abraham'. After Elias came, Elijah, the Messianic forerunner who would come before 'the great and dreadful day of the Lord' (Mal. 4.5-6), appeared and gave to Smith 'the keys of this dispensation', as well as the sealing power to bind and validate all earthly ordinances in heaven (Doctrine and Covenants 110.11-16; see Mt. 18.18).

Smith abandoned the Kirtland Temple when the church moved to Illinois. Interest in Kirtland for religious heritage purposes did not occur until the mid-1900s. However, by then the Community of Christ, comprised of many Church members who did not follow Brigham Young but rather Joseph Smith's son, Joseph Smith III, had purchased the Kirtland Temple.[11] While congregants from both churches engage in what Howlett (2014: 2) calls 'parallel pilgrimage', or 'ritual journeys by disparate groups to a site of some shared superhuman significance', the majority of visitors today are from the Latter-Day Saint Church. However, over time Latter-Day Saint Church leaders have purchased land around the Kirtland Temple and have reconstructed buildings that would have existed at the time the Church was headquartered in Kirtland (Olsen 2004; Staker 2004) to provide historical context for visitors.

Missouri

While in Kirtland, Smith received a number of revelations commanding the main body of the Church to resettle in Missouri. As noted above, other revelations during this time revealed specific locations in Missouri that had close Old Testament and eschatological ties, including the location of the Garden of Eden, Adam-ondi-Ahman, the location of the 'New Jerusalem' and the region where Zion would be built on the American continent. While the Church owns most of land around the area Smith identified as Adam-ondi-Ahman, it is not a well-developed or highly visited religious heritage site. This is due in part because this area was never settled by Latter-Day Saints during the Missouri historical period, and as such does not have the historical resources necessary to attract tourists. As well, Church leaders and members have a hallowed view of this location, and Church leaders would rather rent the land out for agriculture rather than commercialize this location. However, because of this location's importance in Latter-Day Saint eschatology, a few Church members, particuarly those who may hold a literalist interpretation of the book of Genesis, venture to this esoteric spot to see where Adam dwelt and where Christ will again return (Norman 1988), so the Church has installed picnic tables and restrooms at the site.

Nauvoo, Illinois

Religious persecution in Ohio and Missouri again forced Smith to move the church headquarters to Illinois, where he established a city called Nauvoo, meaning 'beautiful' in Hebrew (see Isa. 52.7). Nauvoo was to be a utopian, theocratic society based on city plans Smith had developed in Kirtland in preparation for the creation of the New Jerusalem. It was in Nauvoo, however, where Smith increased his doctrinal expositions related to concepts of salvation, including the relationship between god and humankind, eternal progression, the eternal nature of priesthood covenants, new temple ordinances for the living, and celestial and plural marriage (Lyon 1975). To allow the Saints to participate fully in new priesthood ordinances related to salvation, such as baptisms for the dead, mentioned by Paul in 1 Cor. 15.29, where church members could be vicariously baptized for family members who had not been baptized by proper priesthood authority, Smith built a new temple, known today as the Nauvoo Temple. This practice of baptisms for the dead is an important part of Latter-Day Saint salvation theology and practice and tied to the Church's extensive genealogical program (see below). The Church has purchased much of the land near the Missouri River, as well as the land where the Nauvoo Temple stood,[12] and has since rebuilt the Nauvoo Temple (2002) as well as much of the original city of Nauvoo to show visitors what life was like during this time in church history (Bingham 2002; Olsen and Timothy 2002).

The Mormon Trail

The Mormon Trail is a 1,300-mile trail that was travelled by 70,000–80,000 Mormon emigrants from Illinois to the Salt Lake Valley as they sought refuge after the death of Joseph Smith (see Fig. 4.1). According to Jackson (1992), these Mormon pioneers compared their religious migration to the wanderings of the Children of Israel as they travelled from Egypt to the Promised Land. They also likened the Great Plains to Sinai, the Great Salt Lake in the Salt Lake Valley to the Dead Sea, and Brigham Young, the church leader that led them to the 'Promised Land', as an 'American Moses' (Arrington 1985; Patterson 2015). Church leaders have used the story of the Mormon Trail to create and maintain Mormon historical consciousness through remembering and celebrating the trials of Mormon pioneers as they walked the Mormon Trail (Olsen and Hill 2018). For example, as part of the ritualization of Latter-Day Saint identity, each year thousands of Latter-Day Saint youth in the United States, as well as in distant places such as Australia and Mongolia, re-enact the pioneer trek, mimicking this Moses-like journey of the early Mormon pioneers and reinforcing the biblical connections to this religious migration (Bielo 2016). While this is done in part to 'redeem the dead', one of the missions of the Latter-Day Saints Church (Jones 2006), these treks are designed to encourage youth participants to construct a collective memory between the faithful pioneers of the past and the current generation of Latter-Day Saints, bonding them together through embodied performance and ritualization (Jones 2006; Patterson 2015; Bielo 2016; Olsen and Hill 2018; Hartley-Moore 2020).

Temple Square, Salt Lake City, Utah

The world headquarters for the Latter-Day Saint Church is located in Salt Lake City, Utah. The centerpiece of the 'Church Campus' is Temple Square, which houses the Salt Lake Temple, two visitors' centers and various monuments depicting important historical events of the Restoration. Temple Square serves as a religious heritage tourist attraction, welcoming over five million visitors a year. Because of the interest Latter-Day Saints and non-Mormons have in visiting Temple Square, Church leaders utilize this tourism interest to fulfill its mission to 'teach all nations' (Mt. 28.19) through using the buildings and monuments on Temple Square as pedagogical tools to educate visitors about Latter-Day Saint beliefs (Olsen 2012a). In particular, Church leaders want non-Mormon visitors to leave understanding at a minimum that the Latter-Day Saint Church is a Christian church, that its members believe in and worship Jesus Christ, and that like in the Old Testament, prophets speak for God today (Olsen 2012a, 2019).

To facilitate these interpretational outcomes, the North Visitors' Center houses a number of exhibits that revolve around the Christian nature of the Church and the importance of prophets. On the main floor of the Center, there is an interactive replica of the city of Jerusalem from the time of Christ and a series of paintings depicting his ministry. On the top floor of the Center, there are another series of paintings depicting prophets of the Old Testament and an eleven-foot replica of Danish artist Bertel Thorvaldsen's Christus statue (see Fig. 4.2). When giving tours of Temple Square, the tour guides bring visitors to the statue and play a recording of some of Jesus' teachings from the New Testament, and then bear witness of their love for and faith in Jesus and his atoning sacrifice. On the bottom floor of the Center, there are numerous exhibits related to Old Testament prophets, the importance of modern-day revelation, as well as one exhibit describing Jesus Christ as the 'Good Shepherd' who ministered to the people in the Old and New World.[13] These displays set the context for an exhibit entitled 'Special Witnesses of Christ', which explains how the Church's prophet and apostles speak for God today and invite all to 'come unto Christ' (Olsen 2019; see Fig. 4.3). The South Visitors' Center houses several exhibits related to the construction of the Salt Lake Temple, the importance of families in Latter-Day Saint belief and the purpose of life on earth (Olsen 2012a; Scott 2005).

Figure 4.2. A photograph of the eleven-foot replica of a statue of Jesus Christ by Danish artist Bertel Thorvaldsen, called the Christus. (Source: the authors.)

Figure 4.3. A photograph showing the 'Special Witnesses of Christ' exhibit on the bottom floor of the North Visitors' Center. (Source: the authors.)

Temples and Pageants

In addition to these religious heritage sites, Latter-Day Saint temples and Church-sponsored pageants also encourage pilgrimage-like travel by church members. Latter-Day Saints are a temple-building people. Temples are the holiest spaces in the Church (Jackson and Henrie 1983), and only members who meet certain standards of worthiness may enter. In these temples sacred ordinances of salvation are performed, including baptisms for the dead (1 Cor. 15.29). While there are presently 182 temples around the world in operation, under construction or announced, within the United States there are 81 functioning temples. While travel to these temples is considered a type of 'informal pilgrimage' because of the semi-obligatory and ritualistic nature of temple activities (Olsen 2006), church members generally combine these temple visits with other types of recreational activities when traveling (Olsen and Timothy 2018). Some

Latter-Day Saints have begun to 'collect' temples, in the same way that other tourists collect baseball stadiums or tourism destinations (Timothy 1998; King and Prideaux 2010). Some enterprising entrepreneurs have monetized this desire to collect temples through the creation of 'temple passports', available both in physical form and as an app, which church members can use to keep track of the temples they visit. In addition, in the last few years church leaders have begun to build temples at or near church religious heritage sites as a way of sacralizing these spaces even further (Madsen 2008).

The emphasis on temple work for deceased persons also leads to travel by Church members for genealogical and family history purposes. As a part of the Church's 'work for the dead', the Church has the world's largest genealogy library, and every year thousands of Latter-Day Saint and non-Mormon visitors travel to Salt Lake City to visit the Church's Family History Center (Otterstrom 2008; Ray and McCain 2009). This emphasis on combining religious ritual and obligation with personal heritage goes back to the appearance of Elijah in the Kirtland Temple, where the sealing keys, according to Mal. 4.5-6, would prompt a turning of 'the hearts of the fathers to the children, and the children to the fathers'. Latter-Day Saints interpret this passage to mean that they must 'seek out their dead' and perform vicarious ordinances for them.

Church-sponsored pageants are also an impetus for Latter-Day Saints to travel for religious purposes. Since the 1930s, several large-scale pageants that celebrate Mormon sacred history, including stories from the Book of Mormon, Mormon pioneer journeys West and the life of Jesus Christ, have been founded (Bielo 2016; Bushman 2009; LoMonaco 2009; Hunter 2013; Olsen and Timothy 2018), in part use these 'transformational theatrical ritual[s]' to help build religious identity (Bell 2006; see Jones 2018). Two pageants in particular have strong biblical ties. First, the Hill Cumorah Pageant takes place on the Hill Cumorah, mentioned above. This pageant celebrates the coming forth of the Book of Mormon by the hand of God, and depicts important scenes from storyline of the Book of Mormon, climaxing with the appearance of Jesus Christ in the Americas after his death (see Jn 3.16). Approximately 35,000 people attend this three-week pageant, including church members, interested non-Mormons and anti-Mormon preachers and groups that protest the perceived un-Christian nature of the Church (Gurgel 1976; Olsen 2006; Olsen and Timothy 2018). The second pageant is the Mesa Easter Pageant, with over 80,000 people watching the pageant during its two-week run prior to Easter (Griffiths 2007; Hunter 2013; Jones 2016). This pageant tells the story of Christ's life and ministry through dance and music, with the pinnacle of the pageant being the reenactment of the crucifixion. This pageant is purported to be the largest outdoor Easter pageant in the world (Jones 2016).

Conclusion

The purpose of this chapter was to highlight the interplay between the Latter-Day Saint Church and tourism, with a focus on how the Bible underlies the development of church religious heritage sites in the United States. As noted in this chapter, the literal, dispensational and pre-millennialist reading of the Bible led Joseph Smith to reinterpret the Bible in such a way as to frame the church as a fulfillment of biblical prophecy and a restoration of Christian Primitivism. This view led to many important theological and historical developments and events in Mormonism intimately tied to biblical figures, prophecies and themes. These developments and events have been enshrined in many of the Church's religious heritage sites.

Even though there is no theology of pilgrimage or tourism in the Latter-Day Saint Church, church members engage in informal pilgrimage-like travel to sanctified religious heritage sites related to key figures and events of the Restoration. These places range from where God intervened with human and church history (e.g., Smith's theophany) to more mundane locations devoid of anything but semiotic meaning. The First Vision, as noted above, has led the rise of a small religious tourism niche market revolving around facilitating the movement of church members to sites related to church history and the Holy Land (Olsen and Timothy 2018). While links between Latter-Day Saint tourism and the Bible are clearly made through tours to the Holy Land, the same ties are not always as clear when it comes to travel to church religious heritage sites in the United States. However, as this chapter demonstrates, the Bible has an important and continuing role in the development and maintenance of Mormon historical consciousness and efforts by Church leaders to mark, recreate, and maintain historical sites related to the Restoration.

Notes

1. The Latter-Day Saint Church teaches that following the death of Christ's original apostles, Christianity entered a period of apostasy during which doctrine and texts were altered, de-emphasizing or removing the fullness of the gospel of Jesus Christ from belief and practice. The appearance of God the Father and Jesus Christ to Joseph Smith in 1820 (the 'First Vision') signaled the

beginning of what is commonly referred to in Latter-Day Saint terminology as 'the Restoration of the Gospel', with a reinstatement of the ordinances, offices and doctrine of the primitive church as organized by Christ and the apostles in the first century CE.

2. As part of the Second Great Awakening, several Christian denominations, notably Baptists, Presbyterians and Methodists, were active in western New York in the early nineteenth century. Joseph Smith's mother Lucy, brothers Hyrum and Samuel Harrison, and sister Sophronia, joined the Presbyterian church, while Joseph attended Methodist meetings (see Joseph Smith, *History* 1.5-9). According to Smith, each of the three aforementioned denominations were actively trying to disprove the others, resulting in his queries about which one was 'right.'

3. Moroni was the last prophet to write in the Book of Mormon. He compiled the scriptural writings of his people onto gold plates, which he later presented to Joseph Smith to translate into the Book of Mormon. The account of Moroni's interactions with Joseph Smith (detailed in Joseph Smith, *History*) was canonized as part of The Pearl of Great Price, a compilation of Joseph Smith's writings and interpretation of biblical passages and the Latter-Day Saint Articles of Faith. In addition to the Bible, Book of Mormon, and the Pearl of Great Price, the Latter-Day Saint canon includes the Doctrine and Covenants, a collection of 'revelations given to Joseph Smith, the Prophet, with some additions by his successors in the Presidency of the Church' (Title page, The Doctrine and Covenants).

4. The endowment is a set of rituals in which church members receive blessings and make covenants with God. For more information on the temple endowment, see 'About the Temple Endowment' (https://www.lds.org/temples/what-is-temple-endowment?lang=eng).

5. Smith built two temples during his lifetime – one in Kirtland, Ohio, and one in Nauvoo, Illinois. Smith also dedicated space for additional temples (which were never built) in Independence and Far West, Missouri.

6. Today, the Latter-Day Saint Church has no official position on the location of the Garden of Eden. For the current official statement from the LDS Church, see 'Do Mormons Believe That the Garden of Eden Is in Missouri?' (https://www.mormonnewsroom.org/article/mormonism-101#C18). Brigham Young's teaching on the subject and Wilford Woodruff's recollection are given in Woodruff's journal dated 30 March 1879 (Kenny 1985: 7:129).

7. In the Old Testament, the term 'Zion' referred to certain sacred geographical locations, such as Mount Zion where Solomon built his temple (1 Kgs 8.1; 2 Sam. 5.6-7) and the temple proper (Jer. 31.6). Later the term was used to refer to the city of Jerusalem (Isa. 40.9), the land of Judah (Jer. 31.12), and then the nation of Israel as a whole (Zech. 9.13). In the New Testament, the term Zion was used to refer to the spiritual nature of Christ's kingdom, with Christ as its cornerstone (1 Pet. 2.6). In Latter-Day Saint theology, the concept of

Zion is understood in both a geographic and a spiritual sense, as both where the 'pure in heart' live (Doctrine and Covenants 97.21) and where a City of God or the 'New Jerusalem' will be built on the American continent (Doctrine and Covenants 45.66-67; see also De Pillis 2003). The importance of Zion and the New Jerusalem being located in Jackson County, Missouri, for early Latter-Day Saints is evinced by its inclusion it the Articles of Faith authored by Joseph Smith in 1842: 'We believe in the literal gathering of Israel and in the restoration of the Ten Tribes; that Zion (the New Jerusalem) will be built upon the American continent; that Christ will reign personally upon the earth; and, that the earth will be renewed and receive its paradisiacal glory' (Articles of Faith 1.10).

8. Although more recent LDS Church leaders have taught that Zion should be built wherever the Saints are geographically located, the centrality of Zion in Jackson County, Missouri was still espoused in the mid- to late twentieth century based on Doctrines and Covenants 101.17, 20: 'Zion shall not be moved out of her place...there is none other place appointed than that which I have appointed; neither shall there be any other place appointed than that which I have appointed' (Smith 1956: 72; McConkie 1985: 595).

9. The Joseph Smith Translation of the Bible, sometimes also referred to as the Inspired Version, is a revision of the Bible by Joseph Smith, who believed that many of the 'plain and precious truths' of the Bible had been lost through centuries of copy-editing and translation. He continued to make modifications to his revised biblical manuscript until his death (Matthews 1975).

10. In the Latter-Day Saint Church, the priesthood has the power and authority to act in the name of God. The priesthood has two lines of authority: the Aaronic Priesthood, which deals with the temporal affairs of the Church, and the Melchizedek Priesthood, which deals with 'the spiritual blessings of the church' (Doctrine and Covenants 107.18). Only worthy males are given these priesthoods.

11. After the death of Joseph Smith, there did not seem to be a clear outline of who Smith's successor would be. While most church members followed Brigham Young, the longest serving apostle, to the Utah Valley, other church members followed others who claimed to be Smith's rightful successor. One of these was Joseph Smith III, the son of Joseph Smith. He started the Reorganized Church of Jesus Christ of Latter-Day Saints, later changed to the Community of Christ (McMurray 2004).

12. The Nauvoo Temple was burned down by an arsonist after the Latter-Day Saints left for the Salt Lake Valley.

13. Latter-Day Saints believe that after Christ's ascension into heaven, he visited the peoples of the Americas. This event is the climax of the Book of Mormon (see Jn 10.16).

Bibliography

Anderson, P. L. (1980), 'Heroic Nostalgia: Enshrining the Mormon Past', *Sunstone*, 5 (44): 47–55.
Arrington, L. J. (1972), 'Oliver Cowdery's Kirtland, Ohio, "Sketch Book"', *Brigham Young University Studies*, 12: 410–26.
Arrington, L. J. (1985), *Brigham Young: American Moses*, New York: Alfred A. Knopf.
Bajc, V. (2007), 'Creating Ritual Through Narrative: Place and Performance in Evangelical Protestant Pilgrimage in the Holy Land', *Mobilities*, 2 (3): 395–412.
Barlow, P. L. (2013), *Mormons and the Bible: The Place of the Latter-Day Saints in American Religion*, Oxford: Oxford University Press.
Beal, T. K. (2005), *Roadside Religion: In Search of the Sacred, the Strange, and the Substance of Faith*, Boston: Beacon Press.
Bell, J. A. (2006), *Performing Mormonism: The Hill Cumorah Pageant as Transformational Theatrical Ritual*, PhD diss., Florida State University.
Belnap, D. L. (2011), 'The King James Bible and the Book of Mormon', in K. P. Jackson (ed.), *The King James Bible and the Restoration*, 162–81, Provo, UT: Religious Studies Center, Brigham Young University.
Bielo, J. S. (2016), 'Materializing the Bible: Ethnographic Methods for the Consumption Process', *Practical Matters*, 9: 1–17.
Bielo, J. S. (2017), 'Replication as Religious Practice, Temporality as Religious Problem', *History and Anthropology*, 28 (2): 131–48.
Bielo, J. S. (2018). 'Flower, Soil, Water, Stone: Biblical Landscape Items and Protestant Materiality', *Journal of Material Culture*, 23 (3): 368–87.
Bingham, J. H. (2002), *'Packaging the 'Williamsburg of the Midwest': Nauvoo, Illinois, 1950–2000*, PhD diss., University of Illinois at Urbana-Champaign, Chicago.
Brayley, R. E. (2010), 'Managing Sacred Sites for Tourism: A Case Study of Visitor Facilities in Palmyra, New York', *Tourism*, 58 (3): 289–300.
Bushman, C. (2009), 'The Pageant People: A Latter-Day Saints Appropriation of an Art Form', in H. du Toit (ed.), *Pageants and Processions: Images and Idiom as Spectacle*, 217–23, Tyne: Cambridge Scholars Publishing.
Bushman, R. L. (1984), *Joseph Smith and the Beginnings of Mormonism*, Chicago: University of Chicago Press.
Davies, D. J. (2000), *The Mormon Culture of Salvation*, Aldershot, UK: Ashgate.
De Pillis, M. S. (2003), 'Christ Comes to Jackson County: The Mormon City of Zion and its Consequences', *The John Whitmer Historical Association Journal*, 23: 21–44.
Erekson, K. A. (2005), 'From Missionary Resort to Memorial Farm: Commemoration and Capitalism at the Birthplace of Joseph Smith, 1905–1925', *Mormon Historical Studies*, 6: 69–100.
Galbraith, D. B., D. K. Ogden and A. C. Skinner (1996), *Jerusalem: The Eternal City*, Salt Lake City: Deseret Book Company.
Griffiths, L. (2007), 'Mesa Mormon Temple Prepares for Easter Pageant', *East Valley Tribune*, 24 March, http://www.eastvalleytribune.com/get_out/mesa-mormon-temple-prepares-for-easter-pageant/article_2a9e847e-9dae-5aa2-923b-66621fd1cc10.html (accessed 22 March 2018).
Gurgel, K. D. (1976), 'Travel Patterns of Canadian Visitors to the Mormon Culture Hearth', *The Canadian Geographer*, 20 (4): 405–18.

Hartley-Moore, J. (2020), 'Recreating the Dead: Darkest Tourism and Pilgrimage in Mormon Handcart Pioneer Trek Re-enactments', in D. H. Olsen and M. E. Korstanje (eds), *Dark Tourism and Pilgrimage*, 119–29, Oxfordshire, UK: CABI.

Hedges, A. H., A. D. Smith and R. L. Anderson, eds (2011), *Journals, Volume 2: December 1841–April 1843*, The Joseph Smith Papers 2, ed. Dean C. Jessee, Ronald K. Esplin and Richard Lyman Bushman, Salt Lake City: Church Historian's Press, 2008.

Howlett, D. J. (2014), *Kirtland Temple: The Biography of a Shared Mormon Sacred Space*, Chicago: University of Illinois Press.

Hudman, L. E. and R. H. Jackson (1992), 'Mormon Pilgrimage and Tourism', *Annals of Tourism Research*, 19: 107–21.

Hunter, J. M. (2013), 'Pageants', in J. M. Hunter (ed.), *Mormons and American Popular Culture: The Global Influence of an American Phenomenon*, vol. 2, 166–7, Santa Barbara, CA: Praeger.

Hyde, O. (1842), 'Interesting News from Alexandria and Jerusalem', *Times and Seasons* (3): 739–42, http://www.josephsmithpapers.org/paper-summary/times-and-seasons-1-april-1842/5 (accessed 5 May 2018).

Jackson, R. H. (1992), 'The Mormon Experience: The Plains as Sinai, the Great Salt Lake as the Dead Sea, and the Great Basin as Desert-cum-promised Land', *Journal of Historical Geography*, 18 (1): 41–58.

Jackson, R. H. and R. Henrie (1983), 'Perception of Sacred Space', *Journal of Cultural Geography*, 3 (2): 94–107.

Jones, M. S. (2006), '(Re)living the Pioneer Past: Mormon Youth Handcart Trek Re-enactments', *Theatre Topics*, 16 (2): 113–30.

Jones, M. S. (2016), 'Testimony in the Muscles, in the Body: Proxy Performance at the Mesa Easter Pageant', *Mormon Studies Review*, 3: 11–18.

Jones, M. S. (2018), *Contemporary Mormon Pageantry: Seeking After the Dead*, Ann Arbor: University of Michigan Press.

Kenny, S. G. (1985), *Wilford Woodruff's Journals*, Salt Lake City: Signature Books.

King, L. M. and B. Prideaux (2010), 'Special Interest Tourists Collecting Places and Destinations: A Case Study of Australian World Heritage Sites', *Journal of Vacation Marketing*, 16 (3): 235–47.

Laga, B. (2010), 'In Lieu of History: Mormon Monuments and the Shaping of Memory', *Dialogue: A Journal of Mormon Thought*, 43 (4): 131–53.

Lankford, S. V., R. B. Dieser and G. J. Walker (2006), 'Self-construal and Pilgrimage Travel', *Annals of Tourism Research*, 32 (3): 802–4.

Lloyd, R. S. (2015), 'LDS Apostle Dedicates Newly Developed Priesthood Restoration Site', *Deseret News*, 19 September, https://www.deseretnews.com/article/865637172/LDS-apostle-dedicates-newly-developed-Priesthood-Restoration-Site.html (accessed 27 March 2018).

LoMonaco, M. S. (2009), 'Mormon Pageants as American Historical Performance', *Theatre Symposium*, 17: 69–83.

Lukens-Bull, R. and M. Fafard, (2007), 'Next Year in Orlando: (Re)creating Israel in Christian Zionism', *Journal of Religion & Society*, 9, http://moses.creighton.edu/JRS/pdf/2007-16.pdf.

Lyon, T. E. (1975), 'Doctrinal Development of the Church during the Nauvoo Sojourn, 1839–1846', *Brigham Young University Studies*, 15 (4): 435–46.

MacDowell, B. (1982), 'Religion on the Road: Highway Evangelism and Worship Environments for the Traveler in America', *Journal of American Culture*, 5 (4): 63–73.

Madsen, M. H. (2008), 'The Sanctification of Mormonism's Historical Geography', *Journal of Mormon History*, 34 (2): 228–55.

Matthews, R. J. (1975), *'A Plainer Translation': Joseph Smith's Translation of the Bible, A History and Commentary*, Provo, UT: Brigham Young University Press.

McConkie, B. R. (1985), *A New Witness for the Articles of Faith*, Salt Lake City: Deseret Book.

McMurray, W. G. (2004), 'A "Goodly Heritage" in a Time of Transformation: History and Identity in the Community of Christ', *Journal of Mormon History*, 30 (1): 58–74.

Mitchell, H. J. (2002), 'Postcards from the Edge of History: Narrative and the Sacralisation of Mormon Historical Sites', *Journeys*, 3 (1): 133–57.

Nora, P. (1989), 'Between History and Memory: *Les Lieux de Memoire*', trans. Marc Roudebush, *Representations*, 26: 7–24.

Norman, K. E. (1988), 'Adam's Navel', *Dialogue: A Journal of Mormon Thought*, 21 (2): 81–96.

Olsen, S. L. (2004), 'A History of Restoring Historic Kirtland', *Journal of Mormon History*, 30 (1): 120–8.

Olsen, D. H. (2006), 'Tourism and Informal Pilgrimage among the Latter-Day Saints', in D. J. Timothy and D. H. Olsen (eds), *Tourism, Religion and Spiritual Journeys*, 256–70, London: Routledge.

Olsen, D. H. (2012a), 'Teaching Truth in "Third Space": The Use of Religious History as a Pedagogical Instrument at Temple Square in Salt Lake City, Utah', *Tourism Recreation Research*, 37 (3): 227–37.

Olsen, D. H. (2012b), 'Negotiating Religious Identity at Sacred Sites: A Management Perspective', *Journal of Heritage Tourism*, 7 (4): 359–66.

Olsen, D. H. (2013), 'Touring Sacred History: The Latter-Day Saints and their Historical Sites', in J. M. Hunter (ed.), *Mormons and American Popular Culture: The Global Influence of an American Phenomenon*, 2:225–42, Santa Barbara, CA: Praeger Publishers.

Olsen, D. H. (2015). 'Christian/Atheist Billboard Wars in the United States,', in S. D. Brunn (ed.), *The Changing World Religion Map: Sacred Places, Identities, Practices and Politics*, 3913–26, Berlin: Springer.

Olsen, D. H. (2016), 'The Church of Jesus Christ of Latter-Day Saints, their "Three-fold Mission", and Practical and Pastoral Theology', *Practical Matters*, 9, http://practicalmattersjournal.org/2016/06/29/lds-three-fold-mission/.

Olsen, D. H. (2019), 'Best Practice and Sacred Site Management: The Case of Temple Square in Salt Lake City, Utah', in P. Wiltshier and M. Griffiths (eds), *Managing Religious Tourism*, 68–75, Oxfordshire, UK: CABI.

Olsen, D. H. and J. K. Guelke (2004), 'Spatial Transgression and the BYU Jerusalem Center Controversy', *The Professional Geographer*, 56 (4): 503–15.

Olsen, D. H. and B. Hill (2018), 'Pilgrimage and Identity Along the Mormon Trail', in D. H. Olsen and A. Trono (eds), *Religious Pilgrimage Routes and Trails: Sustainable Development and Management*, 224–46, Oxfordshire, UK: CABI.

Olsen, D. H. and D. J. Timothy (2002), 'Contested Religious Heritage: Differing Views of Mormon Heritage', *Tourism Recreation Research*, 27 (2): 7–15.

Olsen, D. H. and D. J. Timothy (2018), 'Tourism, Salt Lake City, and the Cultural Heritage of Mormonism', in R. Butler and W. Suntikul (eds), *Tourism and Religion: Issues, Trends and Implications*, 250–69, Bristol: Channel View Publications.

Otterstrom, S. M. (2008), 'Genealogy as Religious Ritual: The Doctrine and Practice of Family History in the Church of Jesus Christ of Latter-Day Saints', in D. J. Timothy and J. K. Guelke (eds), *Geography and Genealogy: Locating Personal Pasts*, 137–51, Aldershot, UK: Ashgate.

Patterson, S. M. (2015), 'The Plymouth Rock of the American West: Remembering, Forgetting, and Becoming American in Utah', *Material Religion*, 11 (3): 329–53.

Ray, N. M. and G. McCain (2009), 'Guiding Tourists to their Ancestral Homes', *International Journal of Culture, Tourism and Hospitality Research*, 3 (4): 296–305.

Schott, S. B. (2010), 'Standing Where Your Heroes Stood: Using Historical Tourism to Create American and Religious Identities', *Journal of Mormon History*, 36 (4): 41–66.

Scott, D. W. (2005), 'Re-presenting Mormon History: A Textual Analysis of the Representation of Pioneers and History at Temple Square in Salt Lake City', *Journal of Media and Religion*, 4 (2): 95–110.

Smith, J. (1845), *History, 1838–1856, volume B-1 [1 September 1834–2 November 1838]*, Salt Lake City: Church History Library, http://www.josephsmithpapers.org/paper-summary/history-1838-1856-volume-b-1-1-september-1834-2-november-1838/20.

Smith, J. and R. B. Thompson (1840), 'Minutes and Discourses, Commerce, IL, 6–8 Apr. 1840', *Times and Seasons*, (1): 91–5, http://www.josephsmithpapers.org/paper-summary/minutes-and-discourses-6-8-april-1840/1.

Smith, J. F. (1956), *Doctrines of Salvation: Sermons and Writings of Joseph Fielding Smith*, Salt Lake City: Bookcraft.

Staker, M. L. (2004), 'Historic Kirtland Restoration Completed', *Mormon Historical Studies*, 5 (1): 223–8.

Staker, M. L. (2011), 'Where Was the Aaronic Priesthood Restored? Identifying the Location of John the Baptist's Appearance, May 15, 1829', *Mormon Historical Studies*, 12 (2): 143–59.

Stausberg, M. (2011), *Religion and Tourism: Crossroads, Destinations and Encounters*, London: Routledge.

Stephanson, A. (1995), *Manifest Destiny: American Expansion and the Empire of Right*, New York: Hill & Wang.

Thayne, S. J. (2007), 'In Harmony? Perceptions of Mormonism in Susquehanna, Pennsylvania', *Journal of Mormon History*, 33 (3): 114–51.

Timothy, D. J. (1998), 'Collecting Places: Geodetic Lines in Tourist Space', *Journal of Travel & Tourism Marketing*, 7: 123–9.

Tobler, D. F. and S. G. Ellsworth (1992), 'History, Significance to Latter-Day Saints', in D. H. Ludlow (ed.), *Encyclopedia of Mormonism*, 2:595–8, New York: Macmillan.

Underwood, G. (2005), 'Attempting to Situate Joseph Smith', *Brigham Young University Studies*, 44: 41–52.

Van Dyke, B. G. and L. C. Berrett (2008), 'In the Footsteps of Orson Hyde: Subsequent Dedications of the Holy Land', *Brigham Young University Studies*, 47 (1): 57–93.

Wood, G. S. (1980), 'Evangelical America and Early Mormonism', *New York History*, 61: 359–86.

Yorgason, E. (2010), 'Contests over Latter-Day Space: Mormonism's Role within Evangelical Geopolitics as Seen through Last-days Novels', in T. Sturm and J. Dittmer (eds), *Mapping the End Times: American Evangelical Geopolitics and Apocalyptic Visions*, 49–71, Burlington, VT: Ashgate.

Chapter 5

LOOKING FOR A MIRACLE:
TOURISM, TANYA AND THEURGY AT THE GRAVE
OF THE 'LATE' LUBAVITCHER REBBE

Simon Dein

Introduction

Pilgrimage is a core theme in the anthropology of religion (Morinis 1992; Eade and Sallnow 2013; Turner 1973, 1974a, 1974b; Sinha, Sharma, and Banerjee 2009; Coleman 2014).[1] While the boundaries between pilgrimage and tourism are often blurred, the aim of the person traveling to a sacred site is one way to conceptualize the distinction. Pilgrims hope to attain an emotional and spiritual experience along the way and once they arrive there, while the tourist aims to find pleasure and recreation. However, individual motivations can be multi-dimensional and pilgrimage and tourism are not mutually exclusive.

This chapter will focus upon the *Ohel* – the burial site of the deceased 7th Lubavitcher Rebbe, Menachem Schneerson, who 'died' in 1994.[2] He was held to be *Moshiach* – the Jewish Messiah – by his followers, some of whom maintain that he is still 'alive'. After he died his followers continued to visit his gravesite in the expectation of receiving a miracle. I will provide case studies pertaining to pilgrimage there and discuss them in the wider context of contemporary Lubavitch politics. This chapter contributes to the existing literature on pilgrimage in Judaism and more specifically to the relationship between biblical texts and religious tourism. The Rebbe's teachings and pilgrimage both emphasize the close connection between the physical and spiritual, and pilgrimage to the *Ohel* is seen by Lubavitchers as a form of 'spiritual elevation'. Not only do the miraculous narratives surrounding the dead Rebbe enhance his charisma, but his experienced presence at the *Ohel* strengthens his teachings. Through visiting the *Ohel* visitors are able to actively engage with their theology and cosmology.

I will focus upon a number of issues. In what ways do Lubavitchers interact with these scriptures – in this instance referring to the Rebbe's writings – and what is the relationship between their biblicism and their community practices? Scriptures are not only read, they are also performed in both intersubjective and virtual contexts. Lubavitchers read the Rebbe's writings on a daily basis, but his ideas are also publicly discussed by members of the community. Recently his writings were deployed as a form of bibliomancy. Following his death in 1994 Lubavitchers continued to consult the physical text, his *Igros Kodesh*, to seek the Rebbe's advice, blessings and 'miracles' (Dein 2012). Furthermore, his texts have become artifacts, material objects which signify the Rebbe's presence even post mortem. They ensure an ongoing relationship with him, which due to modern technology occurs both online and offline.

Pilgrimage in Judaism

The topic of pilgrimage in Christianity, Islam and Hinduism has attracted the attention of anthropologists of religion. Work on Jewish pilgrimage is less common, despite the fact that pilgrimage has a long legacy in Judaism. As Theobald (2013) observes, the Jewish tradition of pilgrimage is made problematic by the absence of a central point of pilgrimage, and following the destruction of the Holy Temple in 70 CE under the leadership of Emperor Vespasian, Jews can no longer undertake the journey to appear before the spirit of Yahweh present in the Holy of Holies. This practice commenced following the redaction of the Torah, when Jews were obligated in Halacha 'three times a year all your males shall appear before the Lord your God' (Exod. 23.17). Visiting the Temple was made impossible following the second Jewish revolt under the rule of Hadrian, when the vast majority of Jews were expelled from the Holy Land. The majority of extant work on Jewish pilgrimage has focused upon visits to the graves of saints, the *zaddikim*, particularly in Morocco and Israel (Daryn 1998). Scholarly work on visits to saints' graves in Hasidism has not been accorded the attention received by Moroccan Sephardic *hilluloth*. As we shall discuss later, the phenomenon of miracle-working *zaddikim* also exists among Ashkenazi Hasidim.[3]

Here I specifically focus on pilgrimage to the grave of the deceased Lubavitcher Rebbe as a form of healing ritual along the lines described by Dubisch and Winkelman (2005). They argue, 'Religious pilgrimages and secular pilgrimages allocate the features of the ritual nature of the journey, the power of the special site, connection of the journey to powerful cultural myths, the social and spiritual connections established on the journey, and the transformative nature of the undertaking, including the transformation from illness to health' (xv). For them, pilgrimage includes many facets of healing: 'Pilgrimage is a "multi-media therapy" that combines many different kinds of healing processes that are not normally found within a single healing tradition or activity' (xxxvi). It can be seen as a healing ritual and, as such, results in a radical self-transformation, empowerment, a change in consciousness and a connectedness with the sacred. The self is reintegrated into a collective, in this instance, spiritual model of suffering.

As I shall discuss below, the Rebbe's texts and the narratives of 'miraculous healing' disseminated by Lubavitchers articulate the close relationships between the spiritual and the physical. Pilgrimage to the *Ohel* and the miracle narratives pertaining to this place continuously reinforce the Rebbe's charisma and the spiritual energy found there continues to bring his writings to life. Most importantly, visitors experience the close connection between the physical and divine worlds as the Rebbe discusses in his voluminous writings.

Lubavitch

Hasidism is a form of Ultra-Orthodox Judaism which originated in Eastern Europe and was founded by the Baal Shem Tov. Here I focus on one Hasidic group, Lubavitch (also known as Chabad), comprising about 200,000 members worldwide whose main center is in Brooklyn, New York. Unlike other Hasidic groups, Lubavitch emphasize the teaching of mystical concepts to all its members rather than just a select few. Its mystical text, Tanya, contains teaching from Kabbalah, particularly that of Isaac Luria, the renowned sixteenth-century Kabbalist. The term Kabbalah refers to the inner mystical secrets of the Torah. The Lubavitcher Rebbe was very well versed in mystical ideas, and many of his teachings around Tanya focused on them.

According to Lubavitcher cosmology, the creation of the finite universe, involved God contracting himself into vessels of light (*zimzum*) partly limiting himself to create the world. These subsequently shattered (*Shvirat Keilim*), and their shards became sparks of light trapped within the material of creation. These shards, known as *kelipots*, give rise to evil and personal suffering. Some areas of the universe were illuminated representing goodness, while others remained in darkness representing evil. Prayer and contemplation of various aspects of the divinity (*sephirot*), are able to free these sparks of God's self, thus bringing about a reunion with God's essence, and at the same time bringing them closer to a fixed world. Thus the universe was created in an imperfect state, and the physical and spiritual worlds were forever fused.

Lubavitchers stress the role of prayer, *mitzvot* (good deeds) and ritual in *tikkun olam* (repair of the world). Every Jew is tasked with this process and once completed they will witness the arrival of *Moshiach* (the Jewish Messiah). Lubavitch is well known for sending out *schluchim* (emissaries) to all parts of the world to encourage non-practicing Jews to return to Orthodoxy through the performance of rituals. This is dependent on the view that all Jews possess a mystical spark which can potentially be ignited. Every Jew can elevate him/herself through performing good deeds. The notion of spiritual elevation was commonly deployed by the Rebbe in his voluminous writings and is a core metaphor deployed by Lubavitchers.

The Rebbe

The Hasidic emphasis on the *zaddik* (spiritual leader) differentiates Hasidism from other types of Judaism. Most Hasidic communities narrate miracles that follow a *yechidus* (a spiritual meeting with a *zaddik*): infertile women bear children, individuals with cancer are cured, wayward children become pious, and businessmen are financially successful. Rabbi Menachem Mendel Schneerson assumed leadership of the movement in 1950; he is now simply known as 'the Rebbe' among adherents. Born in Nikolaev, Russia, his followers recount his remarkable abilities as a young child. At the age of two he was already asking the four questions at the Passover Seder. In elementary school he could no longer study with the other children on account of his superior talents. At the age of nine he reputedly saved another child from drowning by jumping into the river and retrieving him. His knowledge of the Torah was held to be outstanding.

During his lifetime, much Lubavitcher discourse centered on the Rebbe, who was seen by his followers as a thaumaturge. They attributed superhuman powers to him, such as the ability to sleep for only two hours a night on a regular basis. His perceived powers were based upon the fact that he was seen as an intermediary between the divine and human worlds, allowing Divine energy to come into the world. Because of his connection to God, he was held to cause blessings to descend from heaven into the world and perform miracles. Lubavitchers saw him as a prophet:

> In our generation, we have merited an individual who meets all the criteria set out by Maimonides: The Lubavitcher Rebbe. The Rebbe's character fits the halachic description of a prophet. 'Great wisdom, a giant in character... his mind is constantly directed towards Heaven'. The Rebbe shared his prophecy with us on numerous occasions, and his words were fulfilled in every respect. The Rebbe predicted the downfall of communism five years before it happened. During the Gulf War of 1991, the Rebbe assured the people of Israel that chemical weapons would not be used, and they would remain safe. The Rebbe was also the first and, at times, the only one to warn the Israeli government of the dangers of land concessions to the Arabs.[4]

Many people had a brief audience with the Rebbe at a weekly ceremony called 'Dollars', at which each person attending would receive a dollar and ask the Rebbe for a blessing. Private *yechidus* with the Rebbe ceased in 1981 and Dollars commenced in 1986. Around 6,000 people were present at this ceremony and Lubavitchers reported miraculous events resulting from the Rebbe's blessings – healing of a relative's sickness, finding a partner, providing infertile couples with children or the acquisition of wealth (Dein 2002). At other times followers would regularly write, fax or email him asking for a blessing hoping for 'miraculous' changes in their life.

Typically, the Rebbe would give the supplicant a blessing and ask them to check their religious artifacts, usually the *mezuzah* or *tefillin*.[5] It was held that any dereliction in these artifacts could lead to physical illness. Repair of these could result in healing of this affliction. Lubavitchers maintain that spiritual disorder can bring about physical disorder; for example, a misspelling of the Hebrew word *Lev* (heart) was held to result in a heart attack (Dein 2002). The Rebbe always asserted the close connection between the physical and spiritual worlds in his teachings. In relation to healing, the Rebbe often stated: 'This is all the more so since spiritual health is generally related to physical health, particularly as far as a Jew is concerned'.

In 1994, he died from a stroke without any successor. For many years his followers had maintained that he was *Moshiach* who would bring in the Redemption. After his death, Lubavitch split into two opposing groups. While some messianists hold that the Rebbe died but is yet to be resurrected, others hold that he is still alive, only concealed. The anti-messianists maintain that the Rebbe could have been *Moshiach* if God had so ordained, but they disagree vehemently that he could return from the dead. The leadership of Lubavitch no longer maintains that he is the Messiah.

Since the Rebbe's death in 1994, his followers continue to visit, write, fax or email his tomb requesting help and advice. His residence, 770 Eastern Parkway, has become a major attraction for Jewish visitors from across the world, with replicas having been built in Israel, Australia, Argentina and Peru. These replicas have become local centers/bases of operations, reflecting the globalizing ambitions of Lubavitch.

There has been an outpouring of his writing over the past twenty years. These include over one hundred volumes of talks presented at *farbrengens* (joyous gatherings), many thousands of his letters published in some thirty volumes, and notebooks found in his private office after his death. The Rebbe spent countless hours over his life teaching the Torah and speaking without notes during the week (when it could be recorded) and also on Shabbat and Festivals (when this was not possible). Over two hundred volumes of his talks, writings, correspondence and responses have been published up till 2020. Topics range from mysticism, Talmud and Hasidic philosophy, to scientific discussions and world affairs, and provide guidance in personal matters, education and social and communal affairs. The Rebbe's talks (*sichot*) and discourses (*maamorim*) make Hasidic and Kabbalistic thought relevant to modern life. One predominant focus in his writings is the close relationship between the physical and spiritual worlds and the importance of ritual performance in cosmic repair.

Men with 'outstanding' memories recorded every word the Rebbe said after the Shabbat or Festival was over and all of this is now printed. Some of his talks were then 'edited' by the Rebbe himself, who added in sources and footnotes. He 'edited' only a small percentage of his talks, so there exists edited talks and non-edited talks in print in different publications. His writings are seen as the Torah and as holy works by Lubavitch.

For his followers – and for many non-Lubavitchers – these posthumously published volumes continue to propagate his teachings. Many who read them are deeply moved by the learning and inspiration they offer. As one Lubavitch rabbi states, 'those ideas are very much what drive the movement today, and having access to them is obviously very crucial'.

Various messianic websites have been developed in recent years to enable connections with the Rebbe. The websites comprise written information, video footage of the Rebbe's *farbrengens* and audio recordings of his numerous discourses. Some of the websites contain video clips of the

Rebbe distributing dollars and sound recordings of his followers singing the *yehi* (a song denoting the fact that he is alive). Autobiographical accounts of individuals whose lives have been significantly influenced by the Rebbe can be found here, emphasizing his miraculous feats pertaining to health, wealth, education and marriage. Many sites provide 'proofs' of the fact that the Rebbe is still alive.

As one illustrative example, www.770live.com provides a live telecast from the 770 location. The site provides Hasidic teachings and includes information on the Messiah, Exile and Redemption and proofs of the Rebbe's messiahship. There is also a discussion of the Rebbe's thoughts on 'current events'. It is also possible to watch a video recording of the Rebbe's synagogue 'live' in real time on this website.

Ohel *Miracles*

Following the burial of Rabbi Yosef Yitzchak Schneerson, the sixth Lubavitcher Rebbe, in the Montefiore cemetery in Cambria Heights in New York in 1950, his successor, Rabbi Menachem Mendel Schneerson, would read out aloud the requests of people who came to consult him, then tear up the notes and leave them at his predecessor's graveside. After the death of his wife in 1988, the *Ohel* (literally, a tent placed over the burial site of a holy person) was the only place the Rebbe regularly visited outside Brooklyn. He suffered his first stroke at the *Ohel* in 1992. The miracle stories continued after Rabbi Schneerson's own death in 1994. But how are his followers able to communicate with the deceased Rebbe? Many Lubavitchers continue to email or fax his gravesite in Queens, whereby his secretary reads out the requests. Others make a pilgrimage to the *Ohel* to have a direct 'audience' there with their dead Rebbe. The site possesses, in Preston's (1992) words, a 'spiritual magnetism' associated with many narratives of miraculous healing there, which in turn add to its spiritual power.

There is a regular bus service running from 770 Crown Heights to the *Ohel*, and passengers are shown videos of the Rebbe distributing dollars en route. On one journey I observed, several Lubavitchers told narratives of their visits to the Rebbe. One man recounted how he received healing for a son who suffered with asthma.

Like other tombs of Jewish, Christian and Muslim saints, the *Ohel* attracts many thousands of visitors a year, who often travel from afar to 'consult' with Rabbi Schneerson, and it has become a major pilgrimage site for Lubavitchers and for those not directly linked to Lubavitch. These are Jews who do not affiliate with Lubavitch but who have generally heard about his 'miracles'. Occasionally Gentiles visit the tomb.[6] Visitors pray to God, request blessings, and connect with the deep spiritual energy of the Rebbe.

Lubavitch explain that visiting the tomb of a *zaddik* is a spiritual act. Those attending become more aware of their limited time on earth. The burial place of a *zaddik* is a holy place and a portal to the heavens. They maintain that *zaddik*'s presence can be felt at his gravesite just as it was felt during his lifetime. As the Zohar teaches us, 'A *tzaddik* who passes on is present even more than in his lifetime'. This itself can bring about a transformation in consciousness. Finally, at the resting place it is easier to connect to the *zaddik*'s soul above and to request his blessings, just as occurred before his passing.

Before approaching the tomb, pilgrims enter a visitors' center where, in accordance with Lubavitch custom, men and women are segregated. The complex is made up of several houses. In the entrance there is a large screen playing videos of the Rebbe's talks with English subtitles. There are many bookshelves with Hasidic books, many of them adaptations of the Rebbe's works. Studying Hasidic teachings is an important preparation for entering the *Ohel*. Tables are set up to write out petitions for the Rebbe. It is customary for male visitors to cover their heads while there and for both sexes not to wear leather shoes. This tradition of taking off leather shoes harkens back to the command to Moses when he approached the Burning Bush (Exod. 3.5), 'Remove your shoes from your feet, for the place on which you stand is holy ground'. People are instructed to walk backwards when leaving the *Ohel* as a sign of respect.

The *Ohel* is closed on all sides but without a roof. Inside there is a second inner wall, waist high, on which people can lean during their prayers. This structure provides a symbolic space, cut off from everyday reality, in which they can commune with the Rebbe. Some Lubavitchers recount how through visiting the *Ohel* they re-experience the 'spiritual elevation' that they felt when they met him while he was alive. Many experience a strong sense of his presence in the sacred space of the *Ohel*. Proximity to the grave is considered meritorious and spiritually beneficial. One Lubavitcher recounted to me:

> The Rebbe lived a life of holiness, a life grounded in spirituality. After his passing, soul-connections continue to form. And for those who study the Rebbe's teachings, visiting the *Ohel* is actually the climax of a continuous relationship with the Rebbe. It's an educational process.

Kvitilim (petitions) over a foot high can be found carpeting the grave. At the *Ohel*, visitors have a tradition of writing *kvitlach* (prayers on small pieces of paper) which are then torn up and tossed onto his grave. In the visitor's center, a fax machine receives over 700 faxes a day, while a computer receives 400 emails daily. These *kvitlach* are all printed and subsequently taken to the graves, where they are torn into shreds and placed atop the graves. Some assert that they have received 'miraculous' responses.

Below I document several instances of 'miraculous' healing occurring at the *Ohel*. Each documents the healing power of the Rebbe as being above and beyond that of doctors. These narratives are ideological in portraying the ability of the Rebbe and the subsequent enhanced faith of the narrator or of the other characters involved in the stories.

Healing One

Mordechai travelled from the United Kingdom to visit the Rebbe's tomb. Born into a Lubavitch family in Stamford Hill, London, Mordechai had spent many years teaching in a Jewish boy's school. Married with nine children, his mother was seriously ill with bone cancer. When visiting the *Ohel*, he petitioned the Rebbe to provide a cure for her and give her the ability to get through her chemotherapy. During my interview with him several months after he had returned to Britain, he recounted that his mother had gone into remission and was functioning well. He impressed upon me the fact that despite his 'apparent death', the Rebbe is still very active in the world. Mordechai was a messianist, maintaining that the Rebbe was not only the Messiah but that he had never died. Again, he stressed the fact that the Rebbe had intervened in his mother's healing. Unlike the other stories cited below, he did acknowledge that chemotherapy had in fact helped his mother.

Healing Two

This second instance relates to infertility.[7] It describes how the power of the doctors, in this instance to cure infertility, pales into significance compared to that of the Rebbe. Reuven, a Chilean doctor, recounted the following story about getting lost on his way home:

> My wife and I had two beautiful children, and we wanted more. We waited and prayed, to no avail. After nine or so years, we visited the best doctor in this field, who told us to be happy with the children we have, as there was no way we would have any more. I go to work in Manhattan every morning by train. One day, I decided to take my car. On my way home, I encountered

a heavy rain. Not being well-acquainted with the route, and with limited vision, I lost my sense of direction. I drove around in circles and could not find my way. Eventually, I recognized a street name – Francis Lewis Boulevard. I believed this street would lead me to my neighborhood, so I continued on it until I came to a dead end. I had to go right or left. I had no idea which way to go.

By chance he encountered a Chabad House and a Lubavitcher there gave him directions to the *Ohel*.

> In response to my request for directions, the *chossid* inside handed me a slip of paper. Back in my car, I looked at the paper and noticed that it had directions listed to all the different highways. And then I saw the word – *Ohel*. Oh, this is the *Ohel*! I had never visited the *Ohel*, and never intended to visit to the *Ohel*. But there I was on a cold and rainy night, sitting in my car near the *Ohel*. If Hashem brought me here, I thought to myself, maybe I should go in. I got back out of my car, knocked on the door again, and asked about the procedures of davening at the *Ohel*. The chossid explained what to do. I went out into the rain and stood at the *Ohel* for quite a while, all alone in the quiet of the night, and I davened to Hashem in this very holy place, the resting place of a great *tzadik*, for everything I could think of. Six months later my wife was pregnant, and on Erev Shevuos, 5770, we were blessed with a baby boy. We felt that Dovid would be the perfect name for a Shevuos baby, and we also felt blessed with this wonderful *brachah*, so we named our son Dovid Boruch. We were very grateful and happy beyond words. Gimmel Tammuz approached, and we were encouraged to return to the *Ohel* to thank the Rebbe for interceding for us with Hashem, who gave us the wonderful blessing of our son.

Reuven is now a frequent visitor to the *Ohel*. He is learning Tanya along with his other Torah studies, and he feels that miracles like this should be publicized.

In this instance, a woman was infertile for nine years having visited the best doctors available. It was only when serendipitously her husband lost his way, was directed to the *Ohel*, and prayed over (to?) the deceased Rebbe that she became pregnant. He maintained that God had directed him there. There is no information provided as to the medical cause of this infertility and in this story doctors are dismissed in one line, only to prioritize the miraculous feats of the Lubavitcher Rebbe.

Healing Three

In the next example, the sick person made a full recovery following a visit to the *Ohel*. The way the story is recounted privileges certain cultural themes (e.g., the Rebbe's power) and suppresses others through its depiction of characters in the miraculous healing story. The events unfold to exclude any possibility of healing apart from the Rebbe's intervention.

> My then-28-year-old daughter took a trip to Israel with an extended stopover in Spain. Arriving in Israel, she noticed swelling in her right arm, which she attributed to a bug bite or something she ate. Naturally, she didn't give it much thought. But the swelling persisted, and on the Sunday she returned to the United States, she called my cousin, a pediatric oncologist. After hearing the symptoms, he came to the conclusion that it must be a blood clot and advised her to go immediately to the hospital. Living in Manhattan, she made her way to the New York University Hospital emergency room, where an ultrasound scan confirmed the clot. Strangely enough, her condition was one that usually appears as a result of activities that demand intensive straining of the arm, such as baseball-pitching or weight-lifting – neither of which my daughter had ever attempted. When she notified me that she was on her way to the hospital, I dropped everything to meet her there. The doctors decided on a rather simple operation that entailed inserting a tube directly into the clot, through which a dissolvent medicine would be injected. This would be followed by a CT scan to determine if the blood clot had dissipated. It all sounded rather simple. The procedure took place on Tuesday. 'When I went to visit her in the recovery room following the operation, she was suddenly overcome with an intense, excruciating pain in her abdomen and back, causing her body to spasm uncontrollably. The doctors injected her with painkillers, and after the pain subsided, she was transferred to the intensive-care unit. The tube that was inserted during the procedure needed to be re-opened, as the short passage of time already caused the slot for the dissolvent medicine to close up. I slept in the hospital that night. The next morning, we were notified that the treatment for the blood clot would be put on hold because, for no apparent reason, my daughter's kidneys abruptly stopped functioning. The doctors were baffled as to what had caused the kidney failure, and very soon, every department became involved in her case. She underwent numerous tests, but none of the results pointed to anything that could be deemed the source. In the meantime, the doctors began to drip liquids into her body to entice the kidneys to begin working again; over time, her body became bloated from the accumulation of liquids. At the end of the week, they began dialysis in the hope of at least cleaning out the poisons from her blood stream. A week-and-a-half on dialysis brought no improvement in her condition. Her situation began to seem hopeless and never-ending. That's when I received a call from Rabbi Yitzchak Weber, the Chabad-Lubavitch emissary in my area. He had heard of our situation and offered to go with me to the *Ohel* in Queens, NY, where I could write to the Lubavitcher Rebbe, Rabbi Menachem M. Schneerson,

> for a blessing for my daughter's recovery. I wasn't the biggest believer, but I figured it couldn't hurt. 'We left the hospital together at 10 p.m., arriving at the *Ohel* after midnight. I wrote my request, and upon the advice of Rabbi Weber resolved to begin laying *tefillin* twice a week. That Shabbat, it was decided that my daughter would be transferred to the dialysis department where, in addition to convenience, she would also avoid the risk of contracting any of the diseases that might have been more prevalent in the ICU. On Sunday morning, after two weeks of endless tests, dialysis and IV drips, my daughter began showing signs of recovery. She went to the bathroom for the first time in weeks, a clear indication that her kidneys had begun functioning once again, just as suddenly as they had collapsed two-and-a-half weeks earlier. Over the course of the following week, she released 28 liters of fluid that her body had accumulated from two weeks of her kidneys not functioning. She was soon back to normal, and upon being released from the hospital, the doctors prescribed oral blood-thinners. Nine months later, the blood clot disappeared. There is no evident medical explanation for all that happened – not for her kidney failure, nor for her sudden recovery. Although the doctors remain mystified by this medical mystery, I am not, considering the fact that her recovery took place just two days following my visit to the *Ohel*.[8]

Like many stories of religious healing, the above story stresses that the 'cure' resulted from religious rather than biomedical means. It was the Rebbe rather than the doctors who was ultimately able to help the sick woman. In this instance the doctors were portrayed as relatively impotent compared to the Rebbe. The narrator becomes connected to the Rebbe, his faith in him increases, and so too his religious observance. Thus, the healing is purposive in increasing the narrator's Jewish religiosity.

Healing Four

In this final example, a man recalls early bruising and interprets it as a form of divine providence prompting a visit to the doctor. Following a diagnosis of cancer, the supplicant was instructed to ask for a miracle which subsequently resulted in healing. The Lubavitcher rabbi requested that he should ask for a 'miracle' rather than a complete recovery, signifying the Rebbe's great powers.

> In January 2015 (5775), I was sitting in my office when I suddenly sneezed very hard. I felt a sharp pain on my left side, and I attributed it to a pulled muscle. Although I was in a lot of pain, I felt better after a few days. Shortly afterwards, I was again sitting in my office when strangely enough, the scene repeated itself. The pain wasn't as intense as in the first instance, but the next morning, I noticed a huge bruise on my left side. I visited my doctor,

who sent me straight to the hospital on suspicion of internal bleeding. After running some scans and tests, the doctors determined that the bruising was a result of an artery I must have torn when I sneezed. Alarmingly, the scans pointed to something much greater that had no connection to the bruising. There was an 8.7-centimeter growth on my left kidney that was releasing blood as well. In retrospect, the sneezing and bruising were actually an act of Divine providence, as they prompted me to have myself examined, thereby uncovering the much greater issue. After three days in the hospital, the doctors informed me that the bleeding had stopped. Though I was being released, they advised me to consult with a nephrologist right away. Upon consulting with one doctor, I was told that no biopsy was necessary, as there was little chance that the growth was not cancerous, and that an operation was necessary to remove most, if not all, of the kidney. I asked the doctor if the surgery could wait until after my vacation scheduled a couple months later, and he assured me it was OK. Nevertheless, I sought out the opinions of two more specialists – one from the University of Pennsylvania and the other from the Fox Chase Cancer Center. Both concurred that the procedure should take place as soon as possible. I decided to undergo the operation with the Fox Chase medical team, and we scheduled for Feb. 12 (Tevet 21). Rabbi Weber, with whom I had grown closer with as a result of my daughter's situation, suggested we go visit the *Ohel* and ask for a blessing that all should go well. Once there, I wrote my note, in which I asked the Rebbe for a blessing for a complete recovery. In the days leading up to my operation, I underwent various tests to monitor the growth in my kidney, and it was quickly decided that its full removal would be fine, as my other kidney was in full working order. At one point, the doctor administered a chest scan as well. The results proved to be terrifying. Two lymph nodes – the size of 3.5 centimeters and 3 centimeters – were detected behind my trachea. This looked very suspicious, as it was the exact area to where the cancer from the kidney was most expected to spread. If this was the case, then the disease was already in stage four. At first, the doctor was hesitant to specify the implication of this, but upon my insistence for a clear prognosis, he informed me that the average life expectancy of a stage 4 patient was less than three years. That night, Rabbi Weber called, as he often did, to hear an update of my situation. When I shared the grim prognosis, he had one question for me: What exactly had I written in my letter to the Rebbe? When I told him I had simply asked for a complete recovery, he was surprised. "You should have asked for a miracle," he gently chided me. He suggested that I write another note with that request, which he then sent with an acquaintance of his to be placed at the *Ohel*. The day of my operation arrived. The surgery stretched on for more than six hours – much longer than expected. Afterwards, the doctor spoke to my family, explaining that everything about the kidney they had extracted screamed cancer (it was sent for further testing to confirm the expected and seemingly obvious diagnosis). If confirmed, I would have to undergo an intensive and lengthy treatment to battle the disease…and we could only hope for the best. After five days of anxious waiting for the lab results, my doctor came back with unbelievable news. The growth was benign! He had made sure that the head pathologist himself thoroughly examined the kidney, and lo and behold, not one cancer cell was detected. I must tell you that when I chose my doctor for this procedure, I made certain to choose the most experienced and acclaimed. My doctor had personally performed 6,000 similar procedures, and his medical group at Fox Chase had completed more than 15,000 such operations. From all these cases, he said he had never seen an instance similar to mine. The size, texture, look and composition of the growth shouted cancer, but the tests have proven it to be completely clean! That Thursday, I underwent a biopsy to determine whether the lymph nodes detected in my chest scan were infected. Following the test, I went home to await the results and was feeling quite optimistic. I had recovered very well from the surgery the week before, and I felt strong and healthy. The next day, a friend came to visit me at home, and as I put up the tea kettle to boil, I was suddenly attacked by tremendous pain. All at once, my hearing dulled, my speech became slurred, and I couldn't stand on my feet. I was rushed to the nearest hospital, where it was determined that I had suffered three successive mini-strokes. Astonishingly, I recovered quickly from this as well; within a week, I was back home. Soon thereafter, the results of my biopsy came back clean. The only possible explanation my doctor managed to come up with was that I had truly contracted the disease, and in some inexplicable way, my body had absorbed it – a medical phenomenon that defies comprehension. At that point, I was not even surprised, as it was evident that I was the beneficiary of extraordinary blessings. Every detail from start to finish was truly a part of this miraculous tapestry of events, beginning with the sneezing and bruising, which, having no connection to my kidney disease, merely served as a warning signal for me to have myself examined. It all ended with a clean bill of health, despite the grim prognosis of the doctors. Contemplating all this, I was stunned by a recollection that still makes my hair stand on end every time I think about it. When I had asked the Rebbe for a blessing for my daughter half-a-year earlier following her kidney failure, I had written that, if necessary, I was ready to sacrifice myself and take her place. Today, I know that one must not ask for such things. Soon after, I made a dinner to give thanks and celebrate my miracle, where I announced that I would be traveling to the *Ohel* to give thanks and express my gratitude, and I urged that anybody in need of a blessing should join me. I rented a limo bus, and we filled it with people. Since then, I have arranged regular trips to the *Ohel*, accompanying and assisting first-time visitors. My parents are Holocaust survivors, and my father became very anti-religious as a result of his experiences. This is the type of home that I was raised in. Without a doubt, the episodes recounted above have completely changed my outlook on life, imbuing me with an entirely new appreciation and sense for true fulfillment. I now lay *tefillin* every day, attend shul on Shabbat, and have fully dedicated myself to assisting Rabbi Weber in building Chabad in our community. And that is the recap of two different stories that wound up having happy endings.[9]

Like Healings Two and Three above, the healing was held to be unlikely in biomedical terms (with a very low likelihood of the original tumor being absorbed by the body). The narrator attempts to persuade us that the cure was really brought about by the Rebbe rather than the biomedical treatments. This signifies to the audience that something impossible did take place, that a person was indeed healed miraculously. Faith in the Rebbe's ability is enhanced as well as the level of the narrator's religious observance. Finally, even though the supplicant had grown up in a secular household, he was keen to point out that this experience had significantly changed his outlook on religious practice.

Conclusion

There is a longstanding pilgrimage tradition in Lubavitch. Before the Rebbe's death his followers would travel to his home in Brooklyn to petition him and receive blessings. Following his death they subsequently traveled to his tomb in Queens. With modern technology it is now possible to send a request via the internet.

The narratives provided here are ideological in portraying the ability of the Rebbe and the subsequent enhanced faith of the narrator or of the other characters involved in the stories. The Rebbe is portrayed as an effective healer in circumstances in which the doctors healing has not been successful. They confirm the view of Dubisch and Winkelman (2005) who stated that 'life transforming experiences are at the core of both traditional and more contemporary forms of pilgrimage' (ix–xxxvi). Not only do supplicants report changes in self-identity, empowerment and physical healing but they also intensify their levels of religious observance like laying *tefillen*.

As Cantwell Smith (1993) notes, scriptures are dependent upon communities of practice to recognize them as such. In order for a text to be scriptural it must constantly be endowed with the appropriate significance by a defined group of interlocutors (Bielo 2009). For Lubavitchers the Rebbe's texts are authoritative and they provide a cosmology and moral code for leading their lives and articulate the close relationship between the spiritual and physical.

But how do these acts of pilgrimage enrich the possibility that the Rebbe's words can become scripture for Lubavitch? Not only does pilgrimage to the *Ohel* and the miracle narratives pertaining to this place continuously memorialize the Rebbe and reinforce his charisma, but also the spiritual energy present there continues to bring his writings to life. Furthermore, pilgrimage to the *Ohel* is a form of legitimation of the Rebbe's writings, thus bolstering the authoritative force of his scriptural canon.

We might argue that since the time of his death the Rebbe lived on through his writings, they became his body and this body became vitalized at the *Ohel*. Most informants stated how these stories brought them closer to the Rebbe and reinforced their faith in his powers. One informant referred to his visits there as 'recharging his spiritual battery'. Informants spoke about re-experiencing a state of 'spiritual elevation', just as they did while he was alive. Most importantly visitors experience the close connection between the physical and spiritual worlds as the Rebbe discusses in his writings. Through visiting the *Ohel* visitors are able to actively engage with their theology and cosmology.

Pilgrimage is, as Mircea Eliade (1959) suggests, a rite of passage from the profane to the sacred, a sacred place of communion with the transcendent. Coleman and Eade (2004) assert that pilgrimage is a performative action that acts as an analogy for other activities or ideas that do not actually involve physical displacement. In this instance pilgrimage is a metaphor for elevation of the divine sparks. This perspective focuses upon embodied experience during ritual. I suggest there is an analogy between the Rebbe's spiritual writings and pilgrimage, both of which stress the importance of elevation.

Notes

1. Anthropological study of pilgrimage has increased in recent years, and much work has been dominated by one of two theoretical paradigms. The older, Turnerian depiction of anti-structure and *communitas* has been influential but appears to be contradicted by Eade and Sallnow's more recent emphasis on the sacred as 'contested' at the great (Christian) pilgrimage sites. Recent social science scholarship on Lubavitch has focused on two themes: the resolution of cognitive dissonance following his death and the role of visual culture in rendering the dead Rebbe present. In relation to dissonance (Dein 2010, 2011), I have looked at the rationalizations deployed to allay cognitive dissonance in the wake of the failed prophecy, particularly that of spiritualization. In relation to visual culture, Shandler (2009) has examined the use of visual media in creating a virtual Rebbe. Maya Balakirsky Katz describes the visual culture of

2. Chabad: the vast and complex visual tradition produced, revered, preserved, banned, and destroyed by the Hasidic movement of Chabad. This rich material culture includes the hand-held portrait, the 'rebbishe' space, the printer's mark, and the public menorah.

2. I have conducted ethnographic fieldwork among Chabad communities in Stamford Hill (United Kingdom) from 1989–2016. I lived with a Lubavitch family and attended Lubavitch teaching sessions, Jewish festivals, and Lubavitch celebrations. I attended a number of Tanya sessions and read the texts and discussed various interpretations with them. I also read numerous books describing the Rebbe's life and various magazines published by Chabad describing the Rebbe's miraculous feats. I visited Crown Heights (Brooklyn, New York) on three occasions, each for a period of one month in 1993, 1995 and 2013, which allowed me to obtain a diachronic view of messianic beliefs. I visited the *Ohel* on several occasions after the Rebbe's death.

3. The spread of Jews into the Diaspora resulted in the development of local pilgrimages to the graves of saints (*Hilluloth*). These local practices among Sephardic Jews can be seen as reflecting cultural syncretism between Judaism and Islam. The large-scale immigration of Moroccan Jews to Israel post-1948 saw these practices flourishing there; local saints were created to satisfy the religious needs of the local community. Daryn (1998: 352) states: 'Veneration of saints plays a central role in Moroccan Jewish life and is an important component of their ethnic identity' (cf. Ben-Ami 1984: 207–13; Bilu 1984: 44). The *tzaddik* is a pious man, well versed in the Torah and kabbalistic studies, charismatic and spiritually compelling (Bilu 1984: 44; Ben-Ari and Bilu 1987: 285), deeply religious, honest and innocent. As a saint, he possesses supernatural powers which enable him to influence events and people and to intercede with God on their behalf (Weingrod 1990: 13). These powers do not fade when the saint dies and can continue to benefit his adherents (Ben-Ari and Bilu 1987: 285). Among Moroccan Jews, faith in saints is strongly entwined with the Jewish mystical tradition and with the Maraboutistic element that characterizes North African Islam (Goldberg 1983: 67–8). However, unlike Islam, Jewish custom centered around the graves of deceased saints (Bilu 1987: 285; Ben-Ami 1984: 190), although there were also living saints in Morocco (Ben-Ami 1984: 46–55). What differentiated these *tzaddikim* from ordinary mortals was their ability to perform miracles; to cure the sick, eliminate danger, protect and rescue (Weingrod 1990: 13–14). A person who had received a miracle often became the saint's 'slave'; that is, a special relationship developed between the saint and the 'slave' who submitted completely to the saint and accepted his every pronouncement (Ben-Ami 1984: 54–5). It is now even possible to perform virtual pilgrimage to Jewish Moroccan cemeteries (Boum 2018). Several websites contain photographs of cemeteries where ancestral tombs can be visited in cyberspace. Aside from Morocco, Tunisian Jews and their French and Israeli descendants continue to visit the El Ghriba synagogue on the island of Djerba (called *Ziara*) during the *L'ag Ba 'omer* holiday, a minor festival taking place in May (Deshen 1997).

4. Source: www.southbrunswickchabad.com/page.asp?pageID=%7BE2FDD950-B1AE-45DD-A190-F6088886869C%7D&displayAll=1 (accessed 4 March 2020).
5. The *mezuzah* is a parchment inscribed with verses from Deuteronomy inside a decorative case. *Tefillen* are phylacteries.
6. According to one report British supermodel Naomi Campbell was seen praying at his tomb, www.foxnews.com/entertainment/naomi-campbell-prays-lubavitcher-rebbe-grave (4 March 2020).
7. Source: www.shmais.com/chabad-news/latest/item/miracle-the-ohel (accessed 4 March 2020).
8. Source: www.chabad.org/library/article_cdo/aid/3555176/jewish/The-Ohel-and-My-Medical-Miracles.htm (accessed 4 March 2020).
9. Source: www.chabad.org/library/article_cdo/aid/3555176/jewish/The-Ohel-and-My-Medical-Miracles.htm (accessed 4 March 2020).

Bibliography

Balakirsky Katz, M. (2011), *The Visual Culture of Chabad*, Cambridge: Cambridge University Press.

Ben-Ami, I. (1984), *Saint Veneration Among the Jews in Morocco*, Jerusalem: The Magnes Press.

Ben-Ari, E. and Y. Bilu (1987), 'Saint Sanctuaries in Israeli Development Towns: On a Mechanism of Urban Transformation', *Urban Anthropology*, 16(2): 243–72. Reprinted in E. Ben-Ari and Y. Bilu (eds), *Grasping Land*, 61–84, Albany, NY: SUNY Press.

Bielo, J., ed. (2009), *The Social Life of Scriptures: Cross-Cultural Perspectives on Biblicism*, New Brunswick: Rutgers University Press.

Bilu, Y. (1984), 'The Folk-Veneration of Saints Among Moroccan Jews in Israel – Forms and Meanings' (Hebrew), in *New Directions in the Study of Ethnic Problems*, 44–50, Jerusalem Institute for Israel Studies, 8, Jerusalem: Jerusalem Institute for Israel Studies.

Boum, A. (2018), *Mémoires de l'absence: les Juifs vus par les musulmans au Maroc*, Rabat: Université Internationale de Rabat.

Cantwell Smith, W. (1993), *What Is Scripture?*, Minneapolis: Fortress Press.

Coleman, S. and J. Eade, eds (2004), *Reframing Pilgrimage: Cultures in Motion*. London: Routledge.

Daryn, G. (1998), 'Moroccan Hassidism: The Chavrei Habakuk Community and its Veneration of Saints', *Ethnology*, 37 (4): 351–72.

Dein, S. (2002), 'The Power of Words: Healing Narratives among Lubavitcher Hasidim', *Medical Anthropology Quarterly*, 16 (1): 41–62.

Dein, S. (2010), 'A Messiah from the Dead: Cultural Performance in Lubavitcher Messianism', *Social Compass*, 57 (4): 537–54.

Dein, S. (2011), *Lubavitcher Messianism: What Really Happens When Prophecy Fails*, London: Continuum.

Dein, S. (2012), 'Internet Mediated Miracles: The Lubavitcher Rebbe's Online Igros Kodesh', *The Jewish Journal of Sociology*, 54: 27–45.

Dubisch, J. and M. Winkelman (2005), 'Introduction: The Anthropology of Pilgrimage', in J. Dubisch and M. Winkleman (eds), *Pilgrimage and Healing*, ix–xxxvi, Tucson: The University of Arizona Press.

Eade, J. and M. Sallnow, eds (2013), *Contesting the Sacred: The Anthropology of Christian Pilgrimage*, London: Routledge.

Eliade, M. (1959), *The Sacred and the Profane*, New York: Harper & Brace.

Morinis, A., ed. (1992), Sacred Journeys: *The Anthropology of Pilgrimage*, Westport: Praeger.

Preston, J. (1992), 'Spiritual Magnetism: An Organizing Principle for the Study of Pilgrimage', in A. Morinis (ed.), *Sacred Journeys: The Anthropology of Pilgrimage*, 31–46, Westport: Praeger.

Shandler. J. (2009), *Jews, God, and Videotape: Religion and Media in America*, New York: New York University Press.

Sinha, A. K., K. Sharma, and B. G. Banerjee (2009), *Anthropological Dimensions of Pilgrimage*, New Delhi: Northern Book Centre.

Theobald, S. (2013), 'The *shluchim*, the Rebbe, and the *tiggun olam*: The Two Pilgrimages within the World of the Chabad Lubavitch', in A. Norman (ed.), *Journeys and Destinations: Studies in Travel, Identity, and Meaning*, 242–64 Cambridge: Cambridge Scholars Press.

Turner, V. (1973), 'The Center Out There: Pilgrim's Goal', *History of Religions*, 12 (3): 191–230.

Turner, V. (1974a), 'Liminal to Liminoid in Play, Flow, and Ritual', *Rice University Studies*, 60 (3): 53–92.

Turner, V. (1974b), *Dramas, Fields and Metaphors: Symbolic Action in Human Society*, Ithaca: Cornell University Press.

Weingrod, A. (1990), *The Saint of Beersheba*, Albany: State University of New York Press.

Chapter 6

Media Pilgrimage:
The Stories that Shape the Modern
Camino de Santiago

Suzanne van der Beek

Introduction

This apostle James is called James the son of Zebedee, or James the brother of John, or Boanerges, which means the Son of Thunder, or James the Greater. (...) after the apostle's death, his disciples, in fear of the Jews, placed his body in a boat at night, embarked with him, although the boat had neither rudder not steersman, and set sail, trusting to the providence of God to determine the place of his burial. And the angels guided the boat to the shores of Galicia in Spain (...) The disciples laid the body of the apostle on a great stone, which immediately softened as if it were wax, and shaped itself into a sarcophagus fitted to his body. (de Voragine 1969: 368–73)

And so the grave of Saint James the Elder was formed, inspiring a pilgrimage that would last over ten centuries. Over time, this story has been told and re-told in various narratives and through different platforms. It is the central legend of the pilgrimage to Santiago de Compostela and has functioned in many ways as the legitimization of the entire ritual. As Simon Coleman and John Elsner put it, legends like these 'reflect the textual recognition of apparent immanence of the sacred, which attracts pilgrims to come to a site' (Coleman and Elsner 2003: 4). This version of the story is taken from one of the earlier texts that invited pilgrims to journey to Santiago de Compostela: *The Golden Legend*. This thirteenth-century collection of hagiographies was one of the most widely read books during the Middle Ages (de Voragine 1969). Although contemporary pilgrims are still aware of this legend, the context in which they encounter it is quite different. They do not encounter it in a collection of stories about the saints, but in the guidebook they have brought on their journey, or in the opening scene of a movie about the Camino, or maybe in a bestselling novel they picked up at the airport. In these contexts, the legend is not presented as an incredible miracle meant to inspire awe and devotion. Rather, it is recounted as a quirky legend, indicating the long history of the cult of the Camino de Santiago.

This is quite exemplary of an important shift that has taken place in the recent history of the Camino de Santiago: the stories that shape the meaning of the Camino have shifted rather radically both in terms of authorship and content. Traditional narratives about the Camino as a ritual of devotion have been largely replaced by media narratives that present a very different interpretation of the journey. This chapter will discuss the Camino as a media pilgrimage by analyzing three of the more popular media stories that inspire contemporary pilgrims on the Camino and exploring the ways these narratives effect processes of identification, expectations and community building. It will argue that in the absence of popular biblical or other explicitly religious stories on the Camino, these narratives effectively inspire a new type of pilgrimage.[1]

Old Pilgrimage, New Pilgrims

The Camino de Santiago refers to a set of roads that leads pilgrims to Santiago de Compostela in the most North-Western region of Spain called Galicia. In recent years, the most route is the 'Camino Francés', which starts in the Pyrenees and travels through the Northern regions of Spain, all the way up to the cathedral of Santiago, where the remains of Saint James the Elder can be venerated. This pilgrimage to this site has been made since the ninth century, though it has had its ups and downs throughout its history (Herwaarden 2002; Costen 1993). The sudden re-popularization of the Camino in the twentieth century was initiated mainly by political processes. General Franco, who had dubbed his Civil War in religious terms as a 'crusade', actively made use of the figure of Saint James to create a symbol for his dictatorial reign of Spain in the first half of the twentieth-century (Talbot 2015). In postwar years, however, the pilgrimage started to be promoted outside of Spain, alongside a number of extensive renovation projects to improve infrastructure and lodgings around Santiago de Compostela (Talbot 2015: 43–4). As the financial benefits of this strategy became evident to an economically faltering Spain, the organization of the pilgrimage opened up to a more international audience. Not long after, European institutions started actively promoting the Camino as a *European* route. In 1985, UNESCO named Santiago de Compostela a World Heritage Site, and two years later, the Council of Europe branded the Camino Francés the First European Cultural Route (Lois González 2013: 13). At this point in time, the religious narrative of the Camino as a Catholic journey of devotion to the Apostle had already been challenged by other narratives: the political discourses of Spanish leaders, the promotion of the Camino as a tourist attraction and the interpretation of the Camino as a European journey.

These stories were clearly told to a willing audience: the number of pilgrims on the Camino has skyrocketed – from 2,491 in 1986 to 277,913 in 2016 (American Pilgrims on the Camino 2017). This new group of pilgrims is not a very homogeneous bunch. Statistics from the pilgrim office in Santiago show that the group includes about as many women as men (52 percent vs. 48 percent in 2016). All adult ages are represented, although the group between 30 and 60 is the largest. About 45 percent of the pilgrims arriving in the cathedral are Spanish, and the other 55 percent consists of a group of pilgrims of 158 different nationalities (Oficina de Acodiga al Peregrino 2017). Although there are no official statistics available on the matter, I have found in my fieldwork that besides pilgrims from Asian countries (mostly South Korea[2]) the great majority of the pilgrims on the Camino are White Westerners.[3]

These contemporary pilgrims are presented with the challenge of finding their own place within a complex dynamic of state and church, leisure and culture, secularity and religion, history and present. Surrounded by a mixed community of pilgrims from all over the world (extending far beyond Europe now), with different backgrounds and with varying degrees of knowledge about the history and political significance of the Camino, contemporary pilgrims asks themselves what the pilgrimage means for them personally within this context. Modern media narratives play a large role in this personal navigation of the Camino, and these narratives have changed quite a bit in shape and content. Whereas we might generalize that for a long time the interpretation of the pilgrimage to Santiago de Compostela was based predominantly on a religious narrative about the devotional journey to the shrine of the apostle, today's Camino stories can cover a large range of different motivations or experiences. This is all the more interesting because many contemporary pilgrims use these new stories to reflect on their own experiences as a pilgrim. In the absence of one defining institutional voice to direct the interpretation of the pilgrimage, these modern narratives become the guiding force in understanding the journey. Pilgrims find these stories via popular media platforms such as Hollywood movies, best seller novels, TV programs, blogs and Facebook groups and they interact with them throughout their engagement with the Camino. These narratives inspire, shape, and mediate their own experiences as pilgrims. Becoming a pilgrim means in many ways to inscribe oneself into pilgrim narratives – appropriating these media narratives means to appropriate the pilgrim identity. In this way, the Camino de Santiago has become a media pilgrimage.

The conception of the Camino as a narratively informed journey is not necessarily a recent construct. In the Middle Ages, hagiographies like the one in the opening of this chapter could inform believers about the significance of the journey to Santiago de Compostela. Biblical narratives also played their role in shaping the form of pilgrimage at that time. In this context, Simon Coleman and John Elsner stress the importance of the story of the two disciples who encounter Jesus after his resurrection on their way to Emmaus, as told in the Gospel of Luke (Coleman and Elsner 2003: 1–16). They argue that '[d]uring the Middle Ages, this story became the biblical model for pilgrimage. Christ himself was perceived to be appearing *as* pilgrim, and was frequently depicted fulfilling such a role in artistic representations of the journey to Emmaus. Here we have a truly scriptural model for the alignment of pilgrimage with the telling of tales' (Coleman and Elsner 2003: 2). However, the interpretation of the Camino as a narratively informed journey takes up a different meaning in the context of a late-modern society in which narratives are produced, shared and consumed at an unparalleled speed and scale. Rather than relating to one authoritative text such as the Bible story recounted above, the contemporary pilgrimage to Santiago is shaped by a large amount of narratives, which are created by different authors, in different cultural contexts, for a wide variety of reasons.

While coming to terms with the Camino de Santiago as a media pilgrimage, we step into a domain that is not often invoked in pilgrimage studies: that of media studies. Broadly speaking, the notion 'media pilgrimage' is a variation on the more widely studied concept of 'media tourism'. Definitions of the concept vary slightly but always revolve around a rather simple idea: media tourism is tourist activity induced by media narratives (e.g., Beeton 2010; Connell 2012). These tourist activities can take a range of different forms, of which one is most obviously recognizable in our pilgrimage site: visitation to locations portrayed in media narratives. Stijn Reijnders, who has worked extensively on the notion of media tourism, refers to these locations as 'places of the imagination'

(Reijnders 2011). He defines this concept as: 'material reference points which have a connection with certain stories' (Reijnders 2011: 14). This concept resonates with the understanding of the Camino as a narratively informed journey. Storytelling plays an important role in the contemporary pilgrimage. The physical engagement with the Camino is almost always preceded by a narrative engagement – there are very few pilgrims who start their pilgrimage without any prior knowledge or fantasies about the journey. As such, the Camino is a place that serves as a materialization of pilgrim narratives and could be considered a 'place of imagination'.

New Pilgrim Stories

When we understand the Camino de Santiago as a narratively informed journey, we should take an interest in the narratives that shape the contemporary pilgrim's vision on that journey. One thing that is easily apparent in studying the more popular Camino narratives is that they are not usually (explicitly) oriented to a religious understanding of the journey. Protagonists might be stated to be Christians, or partake in Catholic rituals along the Camino, but these themes almost never occupy a central position in the story. When claiming that the Camino has developed into a media pilgrimage, we can infer that this process of secularization results in a group of pilgrims that is likewise unconcerned with the journey's religious dimensions. So what stories, then, are being told about the Camino? What picture, if not a religious one, do they paint about the pilgrimage? Let me present an analysis of three popular narratives that inform today's pilgrim community, created in different cultural settings and shared via different media platforms: a Hollywood movie from the United States, a bestselling novel from Germany and a TV series from the Netherlands.

The Way – Finding Yourself by Leaving Your Home

There have been many influential narratives about the pilgrimage to Santiago in recent years. However, years of conversations with pilgrims have shown me that there is one movie that remains the most well-known today. The Camino was still a relatively unknown and untraveled road – especially by those on the other side of the Atlantic – before Emilio Estevez presented his movie *The Way* (2010) in American cinemas. The movie has become a favorite among a worldwide pilgrim community, to the point where pilgrims will sigh that 'the Americans only started coming after *The Way*' (for which there seems to be some actual scientific credence: Lopez, Santomil Mosquera and Lois González 2015). The movie stars Estevez's father, Martin Sheen, whose character Tom travels the Camino de Santiago in order to discover that there is more to life than a career and a steady daily routine. It is a story of a man coming out of his cocoon, reaching out to new environments, new ideologies and a new life.

Tom's journey starts when he hears that his son Daniel has died in the French Pyrenees on his first day on the Camino. Tom and Daniel never really saw eye-to-eye. The neat and orderly Tom could never understand the life choices of his adventurous and free-spirited son. Tom finally leaves his safe environment in order to collect his son's remains. After arriving in France, Tom decides on the spot to have his son's body cremated and walk with his ashes to Santiago. However, Daniel is not the only guide to help Tom in his Camino adventure. During his journey he will meet a set of different people to help him on his way. And he will need them. The movie makes it very clear that Tom is not the kind of man that will be able to fulfill this journey on his own. Before he sets out a French officer asks him if he even knows why he is walking the Camino. When Tom suggests that he is doing it for his son, the officer shakes his head and tells him: 'You walk the Way for yourself, only for yourself'. Tom resigns to his ignorance and replies: 'Well in that case I suppose I don't have a clue!' And sure enough, the moment he walks out of the door of his hotel, he leaves in the wrong direction. However, from then on, he sets out to learn. Accepting that he knows nothing, he is open for help and guidance.

It is significant that Tom had to travel to a remote French village in the Pyrenees in order to start his spiritual learning. In the study of pilgrimage, the notion of leaving familiar surroundings and social structures has played an important part in explaining the spiritual potential of the ritual: by distancing oneself from daily life the pilgrim creates a space of contemplation and introspection. In this condition, one might recognize Victor Turner's famous notion of liminality. Considered in the light of a well-known theory on rites of passage, Turner recognizes that pilgrimage, as a ritual, displays many of its characteristics. Pilgrims can be understood as people passing through undefined places and times in which they are ambiguous, for they 'pass through a cultural realm that has few or none of the attributes of the past or coming state' (Turner 1974: 249). During this uncertain phase, they find themselves 'betwixt and between'. What is more, the reflexive atmosphere that comes about from the pilgrim's separation from home inspires the idea that it is the common surroundings that keep the pilgrim from these moments of insight on a daily basis. It creates the idea that a pilgrim's true self has been pushed away and replaced by short-lived and artificial substitutes generated by a late modern and super diverse society.

In *The Way*, the external forces alienating Tom from his spiritual potential are clearly defined as American. It is therefore not hard to see how the exotic, very non-American environment of the Camino helps Tom to free himself from his daily structures and provide the background for Tom's spiritual awakening. Instead of his rich and well-ordered life in America, where he is constantly called upon by clients, assistants and friends, Tom is now delivered to the lack of modern infrastructure and communication of rural Spain, which forces him to walk on when he is tired, to eat only what is provided to him, to sleep where there is a place left for him. Whenever he relies on his American identity, this only gets him in trouble. Pilgrims jokingly boo him when he is introduced with 'the Americans have arrived', a *hospitalera* shakes her head dismissively when she correctly guess his nationality ('Americans are always late…'), and his American guidebook makes him look a fool when he tries to order *tapa* in a region that traditionally serves *pinchos*. It shows up at its ugliest when Tom has had too much to drink and starts shouting at his friends. When he is arrested, he shouts: 'I am an American, I don't speak Spanish, I speak American. God bless America!' His patriotism is part of him making a fool of himself. It is clear that Tom will need to shed his American-ness in order to succeed on this journey.

By engaging with the Spanish environment, through conversations with his non-American friends and by participating in a European pilgrimage, Tom finally learns to step out of his restrained American way of living. Arriving at the pilgrim office in Santiago de Compostela, Tom is asked for his reason to walk the Camino. Remembering his initial answer given shortly before leaving – 'I am doing it for Daniel' – Tom seems to have realized that he did not, actually, walk the Camino for his son. He now says: 'I guess I needed to travel more'. He has shifted his attention from a fascination with his son's way of life to adopting that lifestyle for himself. What started out as an exotic and unfamiliar journey has now become a very personal one.

This idea of leaving one's physical home to find one's spiritual home has been articulated by Erik Cohen. Cohen argues that pilgrims could be identified by their interest in traveling towards their so-called 'center'. In this, they are fundamentally different from tourists, who are mainly looking to escape their center in favor of the exotic and unknown. In his discussion on 'Pilgrimage and Tourism', Cohen (1992: 50) argues that

> two prototypical, non-instrumental movements can be distinguished: pilgrimage, a movement towards the Center, and travel, a movement in the opposite direction, toward the Other.

The sacred center that attracts pilgrims has been traditionally framed within a religious context, but can now be understood in a broader cultural frame. In Cohen's words, we might say it symbolizes 'ultimate meaning' (Cohen 1979: 181) for the pilgrim. This center, interestingly, lies outside of the pilgrim's daily surroundings: 'The pilgrim, whose ordinary abode is in profane space, ascends, both geographically and spiritually, from the periphery toward the "Center out there"' (Cohen 1992: 51). The world that people inhabit on a daily basis can thus be understood as 'peripheral', while the surroundings of the pilgrimage resonate with pilgrims' true spiritual and cultural lives. In other words: the Camino is a more natural home to pilgrims than their daily surroundings. The very strangeness of the Camino and its pilgrims provide the opportunity for Tom to come closer to himself. It was only in this different location, engaging with people that are completely different from himself, that Tom could find his own spiritual center. And the endurance of that lesson is manifested by the closing scene, in which we see Tom walking in a Moroccan marketplace, dressed in traveler's clothes with the Camino scallop hanging from his neck.

Ich bin dann mal weg – Hape Kerkeling's Many Identities

Among the numbers of people who travel the Camino are a small number of celebrities. One of the more famous pilgrims that completed the journey in the recent past is German comedian Hape Kerkeling. In June/July 2001, he walked the Camino Francés and five years later, his account of the journey was published. Kerkeling's 2006 work *Ich bin dann mal weg* became a big success and sold over five million copies, making it the bestselling German nonfiction book in the last seventy years. In his account of his pilgrimage, Kerkeling focuses on the negotiation between his different identities.

One of the first clues to the main theme of the book is given by the subtitle, which reads: 'Losing and Finding Myself on the Camino de Santiago' (Kerkeling 2009). The whole novel is as much an account of Kerkeling's negotiations with different identities as it is a travel account of a contemporary pilgrim to Santiago de Compostela. Kerkeling sees the task of identity work thrown at him from all different quarters of the Camino: he sees it reflected in billboard advertisements for technological gadgets ('Do you know who you really are?' My answer is quick and clear-cut: "Non, pas du tout!"' [Kerkeling 2009: 7]) and in the questions asked by the officials at the pilgrim office in Saint-Jean-Pied-de-Port. Even before starting the Camino, Kerkeling does not feel himself to have a stable, unified identity. He describes himself as follows:

> Hans Peter Wilhelm Kerkeling, thirty-six years old, Sagittarius, Taurus Ascendant, German, European, adoptive Rhinelander, Westphalian, artist, smoker, dragon (in the Chinese zodiac), swimmer, motorist, utilities customer, TV viewer, comedian, bicyclist, author, voter, fellow citizen, reader, listener, and monsieur. (11)

His identity, then, is a mix of elements relating to different elements of his life. From now on, that of 'pilgrim' will be added to that list. Several times throughout the book, Kerkeling explains that he has always been interested in an open exploration of who he could be, where life could take him and how religion can facilitate the negotiation of these different possibilities. Besides some more practical health problems, this is what led him to the Camino. He hopes that in finding a closer relation to God, he can get a more comprehensive idea of who he himself is. His religious orientation, however, is just as diverse as his secular background:

> I could set off with this question in my head: Is there a God? Or a Yahweh, Shiva, Ganesha, Brahma, Zeus, Ram, Vishnu, Wotan, Manitu, Buddha, Allah, Krishna, Jehovah, etc.? (9)

This openness to the complexity of God(s) mirrors the variety of lessons he develops along the way. The search for those two identities, that of himself and that of God, are closely related during this pilgrimage. As time proceeds, Kerkeling becomes more comfortable with his pilgrim identity. He even proposes that as he is approaching Santiago de Compostela, he is simultaneously drawing nearer to himself.

Although Kerkeling has little regard for his fellow-pilgrims to begin with, the more time he spends on the Camino, the more he starts to identify with the pilgrim identity and with other pilgrims. After about two weeks, however, he starts to reach out to other pilgrims more and more. The first time he realizes that he wants to be in contact with other pilgrims is when he realizes that anyone who has not been a pilgrim will not understand what he is experiencing on the Camino. After barely a week, he comes across a valley where pilgrims over the years have created piles of little stones to visualize their community:

> No one would understand an image like this. If I were to show people a photograph of this valley, they would likely say: 'So, what's that compared to Niagara Falls?' This place imparts strength only to pilgrims, and this valley is special only to them. (59)

At this point he wishes to end his solitude and find a good friend with whom he could talk about his experiences as a pilgrim. The journey makes him look for a connection to other pilgrims, to find common ground rather than fixate on the differences. His identity as a famous German comedian starts to play an important role and it proves to stand in the way of his connection to other pilgrims. It is clear that Kerkeling feels his public identity to be at odds with his newly developing pilgrim identity, and as he is more interested in the latter during his journey, he tries to shun other Germans, who would regard him as a famous comedian rather than a pilgrim. In situations where he comes into contact with his regular audience, he resents them for distancing him from his pilgrim identity, which could undermine his pilgrim identity as well as their own. Resigning to his fate, he steps up to a group of excited fans and thinks: 'Since they have yet to see a vision of the Blessed Virgin Mary on the pilgrim's route, they have to settle for seeing me' (154).

One of the most influential notions used in describing the dynamics of a pilgrim community is that of *communitas*. Victor and Edith Turner, who introduced the term into pilgrimage studies in the last century, describe how being a part of a pilgrim community constructs a unique sense of connection between all participants, which has 'a relational quality of full unmediated communication, even communion' that exist in liminal environments (Turner and Turner 1978: 250). This connection is based on the lack of social structure during a pilgrimage. Every participant has discarded their everyday identity and has thereby also let go of the social position that this identity grants them in everyday society. As a result, many pilgrims remark that interactions on the Camino are more sincere and uncomplicated than in daily life; pilgrims open up to each other with an ease and level of comfort that they do not know at home. Turner and Turner write: 'The bonds of communitas are undifferentiated, egalitarian, direct, extant, nonrational, [and] existential' (250). In Kerkeling's meeting with his fans, it is clear that the famous *communitas* among pilgrims is broken and conventional hierarchical structures are reinforced: they no longer form a group of pilgrims, all equal and united in their shared journey, but have transformed in a group of fans who get a rare opportunity to meet a celebrity. 'Even the American women are ecstatic; they're acting as though they're getting George Clooney in a tux instead of me in my torn denim shirt' (Kerkeling 2009: 154). The dynamic within the group of fans has also changed: the woman who has recognized Kerkeling has risen in rank and now seems to be the leader of the little group. Kerkeling criticizes the group for being impressed with him, for looking up to one pilgrim more than another because he has fame outside of the Camino. 'It shouldn't matter who is standing before you!', he wants to shout at them.

Although this situation excites the fans, Kerkeling is not interested in breaking the pilgrim *communitas*. In discussions with non-German pilgrims, he only brings up his celebrity status as if it were a great secret that he carries around with him. He introduces himself with his full name, Hans Peter, rather than the abbreviated Hape, which he uses in a public capacity, and he only discusses his career with pilgrims who become very good friends with him. And so, his public identity becomes one of his most private secrets, buried well under his pilgrim identity, which he now portrays publicly.

On the other hand, Kerkeling does include his public identity in his own internal reflections on the journey. During one of his first days on the Camino, he tells an extensive story about the long and unsure route to his stardom, which is presented as a parallel to the Camino (Kerkeling 2009: 32–46). He seems to suggest that he is well-equipped to complete the Camino, because he has already completed a similarly arduous and uncertain way:

> These are the kinds of thoughts that are now running through my head. Still, the story I am trying to tell is not the story of my life but the story of the Camino de Santiago – though I've begun to think that in many respects, they're one and the same! (46)

Note that in this quote, he refers to his public career as his 'life'. And so he equates his public identity with his life, and his life with his pilgrimage. All along the Camino, then, Kerkeling makes use of different repertoires to explore his own layered identity: the good and the bad, the public and the private, the secular and the religious. He does not shy away from this complexity, but sees it reflected in the very journey he has started in order to try and find himself.

Op weg naar Santiago de Compostela – The Many Faces of the Pilgrim

Of all the Dutch narratives on the Camino de Santiago, one stands out for its popularity, even outside of the pilgrim community. In 2011, the RKK (the Omroep Rooms-Katholiek Kerkgenootschap, which translates as 'Roman Catholic Church Society Broadcasting Corporation') started to dedicate a series of episodes of their popular program *Kruispunt* to the pilgrimage to Santiago, under the title *De Camino: Op weg naar Santiago de Compostela* (*The Camino: On the Way to Santiago de Compostela*; Kemp 2011–present). This show is an interesting media narrative, because rather than presenting the account of one pilgrim, it presents the stories of many, unconnected pilgrims. The program underlines the variety of stories on the Camino, by including a multitude of voices and by a constant movement between the individual and the collective.

The design of the program invites a closer look at the variation in narratives of different pilgrims. Instead of seeing how one pilgrim narrative develops, we follow presenter Wilfred Kemp, who travels along the Camino and interviews pilgrims he meets. Every episode is built around a selection of portraits, in which every pilgrim displays their interpretation of the journey. Kemp meets among others a blind woman who is led along the Camino by her wife, a grandfather who is mourning his deceased grandchild, a young man who walks 2,700 kilometers to show his family he is as worthy of praise as his brothers, and a woman who reflects on her busy life in a metropolis in the peace of the Spanish countryside. Even though pilgrims tend to talk in broad and abstract terms, Kemp makes a point of asking them to concretize their ideas. For example, many pilgrims initially tell Kemp that they are on the Camino 'to get away from it all' or 'to get to know themselves a bit better'. In these instances, Kemp will ask further: 'What are you trying to get away from?', 'What is it about yourself that you are as yet unfamiliar with?' Via these questions, pilgrims are forced to personalize their stories, and so the diversity that underlies these more general answers becomes apparent. On the other hand, these personal portraits are explicitly embedded in the larger pilgrim community. Visually, the program shifts between close-ups of individual pilgrims and larger shots of the way where we can see groups of pilgrims in similar outfits performing the same walk. Textually, the narrator of the program frames the individual stories with stories about the great number of pilgrims that have traveled this same road for hundreds of years for a variety of reasons and with a variety of backgrounds. Both visually and textually the program zooms in and out of personal narratives, relating individual stories with the larger story of the whole pilgrim community.

This broad interpretation of the pilgrimage and the stories pilgrims create has been an important point of focus for the study of the contemporary Camino. This diversification of pilgrimage has been described using a variety of terms ('liquid', 'fluid', 'networked', 'mobile'), but most apt for this situation might be the idea of elasticity as proposed by Lena Gemzoë (2012). This term 'offers an image of something that can both expand and contrast' and refers to 'it's [pilgrimage's] capacity to absorb or encompass a range of social forms' (41). The term seems appropriate because it suggest both stability and extendibility. Rather than the uprooted connotation of fluidity or liquidity, the notion of elasticity indicates that from a somewhat defined core, the pilgrim narrative can stretch in many different ways. The term 'implies adaptability, which in turn allows for a broader use of the whole concept of pilgrimage, rather than restricting the term to its classic form as purely a religious practice' (41–2). According to Gemzoë, this is an important reason that pilgrimage is such a popular ritual today: it opens up pilgrimage for stories beyond the strictly religious. Kemp stresses this point in the opening of every episode, when he explains that: 'In earlier times, they did this for explicitly religious reasons, to earn indulgences. Today people are more interested in creating a realization of what matters in life.' The diversity of motives among contemporary pilgrims is stressed further by comparing it with those of earlier pilgrims, whose motives are reduced to the obtainment of indulgences. The program shows the pilgrimage to Santiago as a journey that can embrace a variety of stories.

Through the many interviews featured in the program, pilgrims relate their own experiences to that of others on the way. In describing what she has learned on the Camino, one pilgrim says: 'There are so many people here that have a story that is similar like mine, or a little bit different. So you learn that you are not alone.' This pilgrim feels included in the pilgrim community by recognizing herself in other pilgrims' stories. In other interviews, pilgrims look at other pilgrims' stories as points of comparison and use the diversity of stories to reflect on their own unique position.

Throughout the series, the possibilities for variety among pilgrims are played out in different dimensions: diversity in age, religious background, nationality, motivation, social status – they are all well-documented. However, it has also been argued that participating in a pilgrim community leads to a disregard of these differences. We have discussed the notions of liminality and *communitas* before and have seen that these terms are used to refer to a sense of oneness and equality within the pilgrim community. However, the interviews that are conducted by Kemp in *Op weg naar Santiago de Compostela* tend to highlight these differences rather than even them out. Kemp speaks a number of different languages, so that the nationality of the pilgrim that is being interviewed remains apparent throughout the conversation. What is more, Kemp always asks about the pilgrim's lives before they came to the Camino. By doing so, he stretches their Camino narratives beyond the start of the pilgrimage and includes the pilgrim's personal life within it. Every episode is introduced with a short sketch of the history of the Camino and an invitation to listen to 'stories about people who come close to themselves, by leaving their homes'. Again, we find this notion of going back and forth that we have noticed in the zooming in and out of the pilgrim community, although here it is more explicitly related to the pilgrim's life at home and their life on the Camino. Kemp introduces this theme in many of his interviews by asking questions like 'People sometimes say: everyone has their own reason to go on pilgrimage…?'; 'Can you tell a bit more about what made you decide to walk this Camino?'; 'What do you hope the Camino will bring you?'; or simply 'Why did you walk the Camino?' Via these questions, Kemp often stumbles upon personal stories relating to family, work or relationships. Kemp is not the first person to stress this dimension of the pilgrimage. Many studies of contemporary pilgrims show that people are drawn to the Camino at points of their lives when they are faced with some form of personal crisis (Chemin 2012; Frey 1998: 45–6; Gemzoë 2012).

The Camino, then, becomes a place where people come to heal. And Kemp frequently asks whether that actually works – to use the Camino as a cure for whatever personal or spiritual wound someone is walking with. This idea seems to be so prevalent that Kemp anticipates a positive answer to his inquiry into this process. He asks: 'How did the Camino change your life?' (rather than 'Did the Camino change your life?'); or 'What do you take back with you from the Camino?' (rather than 'Do you take anything back with you from the Camino?'). Again, Kemp is not usually satisfied with general remarks that the Camino can have a healing function in a person's life. He will ask pilgrims for concrete lessons they have learned, parts of themselves they have discovered, or parts of their lives that they will change. And pilgrims never disappoint. Every pilgrim presented in the program agrees that the Camino has transformed them, has helped them in a particular way. Even when the change amounts to a stabilization, pilgrims interpret this as a victory.

Due to the format of the TV series, it is possible to bring in a third chapter of the pilgrim narrative: added to the life before and on the Camino, the program shows pilgrims' lives after the Camino. In the second series of the program, Kemp returns to some interesting stories from the year before and finds out how these stories have developed. He visits Marijn, who has fallen in love with another pilgrim on the Camino and consequently left his life as an IT manager in Amsterdam to open up an albergue with her on the Camino. We also hear about the next chapter in the life of a pilgrim whom we met walking the Camino while her husband was ill at home. Her husband had died in the meantime and she has come back to the Camino to end a period of grief.

And so we get an image of the pilgrim story starting a long time before the pilgrim arrives on the Camino and stretching beyond the arrival in Santiago de Compostela, extending into the period after coming home and the different directions these pilgrims' lives take as a result of the Camino.

Many of the pilgrims that Kemp talks to stress that the goal of the Camino is not to reach Santiago de Compostela. Arriving at the cathedral is not the end of the journey, but rather the beginning of a new journey to be continued back home. Only Martijn seems to look upon the journey as a way of proving himself, and clearly walks with a fixed location as his end goal – counting the amounts of kilometers he will have walked when he finally arrives. Still, Kemp asks him about his life after the Camino and he says: 'I know that this is a step in the direction of a better future' (*Op weg naar Santiago de Compostela*, 20 December 2010).

In showing this extended pilgrim story, in which the Camino is presented as a ritual of healing, the program shows pilgrims with an unshakable belief that the Camino does them nothing but good. Even when the entire conversation revolves around physical pain and exhaustion, the conclusions are nothing but positive. For example, in one episode, Kemp speaks with a pilgrim who takes a rest on a bench because her legs are too tired to continue walking. She confesses that the journey is much harder on her body than she had anticipated and we get a shot of the blisters that have formed on her feet. But when Kemp asks her what the most difficult moment was so far, she smiles and says she has had not difficult moments, only beautiful ones. When Kemp asks what her most beautiful moment was, she tells a story about finding a walking stick when she thought she couldn't go on any more. The stick eased her walking and so she could continue. Although her journey is clearly filled with pain and difficulties, she always refers to it as beautiful, rather than difficult. Smiling, she explains, 'Everyone here has pain, it is a part of it'.

Media Pilgrims

The interpretation of the Camino as presented in different media narratives can vary quite a bit. I have argued before that pilgrims use these narratives to navigate the different potential meanings of the Camino. This process of negotiation is played out in different ways. Here, I will discuss three important dynamics that are rooted in the Camino as a media narrative: the identification with narrative protagonists, the pilgrim's positioning between imagination and reality, and the narrative communities that are based on Camino stories.

Becoming a Pilgrim

Going over some of the most popular pilgrim stories, it is hard to say why these narratives specifically appeal to such large groups of people. Reijnders points out that we have no steady research on the types of narratives that induce people to travel to these settings (Reijnders 2009: 168), although suggestions have been made by some. For example, Erik Cohen proposes that people will only travel to the setting of a narrative if that setting is a prominent element in the narrative (Cohen and Cooper 1986). When the setting is of no particular significance or originality, the narrative will not induce mobility. Roger Riley and Carlton van Doren add to this observation that the landscape should be directly connected to the protagonist (Riley and Van Doren 1992). When we look at popular Camino stories, the initial attraction seems to rely mostly on their protagonists. When asked about stories that inspired them to take an interest in the Camino, pilgrims point out that they can relate to the protagonists of these stories. More specifically, they point out that these protagonists do not have any special abilities or knowledge to start a pilgrimage: '…and if they can do it, then so can I!' This remark is often heard when talking about Kerkeling's novel. The public image of Kerkeling is not such that anyone would expect him to complete an 800 km long trail through the North of Spain. In his novel, he does not hesitate to point this out himself: the overweight, non-athletic comedian from Westfalia, stumbling along the Camino without proper preparation. His light-hearted approach to the unsuitability of his own physique takes away any of the self-doubt any reader might have about their own capabilities.

Similarly, in *The Way*, protagonist Tom is not portrayed as the ideal candidate to walk long distances through a foreign country. He is the oldest of his group of fellow-pilgrims, does not speak the local language, has no experience with travel or foreign cultures and has not even prepared for his journey properly. Tom only decides to walk the Camino after he finds out his son has died shortly before he could begin this journey. He decides to use his son's equipment and walk the Camino in his stead. Talking to some of the hospitaleros in the albergue in Roncesvalles (the first resting place for pilgrims that cross the Pyrenees from Saint-Jean-Pied-de-Port) I have been told that since the appearance of *The Way* there has been a significant increase in Americans starting their journey without the necessary preparation: 'They have not broken in their hiking boots. The first time they are walking with their backpack strapped to their backs, they try to cross the Pyrenees!'

In the series of portraits depicted in *Op weg naar Santiago*, we again encounter the idea that anyone could become a pilgrim. Young and old, healthy and ill, athletic or lazy – the series shows the variety of pilgrims and their capabilities. Wilfred Kemp is sure to stress this point in his commentary, and the selection of pilgrims that is portrayed serves to underline this. Perhaps the most striking example is the story of Magda and her blind partner Ineke (*De Camino*, 4 October 2012). The couple has been together for 25 years when they start their Camino to celebrate this anniversary, but Ineke's handicap is obviously an important part of their story. The documentary shows through images and interviews how the two women navigate along the Camino and the difficulties they encounter – and overcome.

All in all, these narratives insist that anyone can be a pilgrim. The identity is not tied up to any type of person, background, capability or motivation. Anyone who takes a fancy can be accepted as a pilgrim. This takes away a large part of the hesitation one might feel when imagining such a journey.

Between Imagination and Reality

Pilgrims feed their interest in the Camino via the variety of Camino narratives available to them. As we have seen, these narratives can offer different themes and motives, among them the exoticism of Northern Spain, conversations with other pilgrims, a search for God or the opportunity to heal. These themes function as pointers that serve as possible elements that pilgrims will engage with during their own journey. By having been included in a famous narrative, these elements become more significant and often function as an interpretive meeting point. Cultural geographer Duncan Light even goes so far as to argue that many travelers are in fact more interested in finding a confirmation of the place they had imagined, than exploring the reality of the actual place:

> In many instances, tourists visit the landscapes represented in a film, rather than the locations where filming took place: what they seek is the place as they want and believe it to be. (...) These are occasions for dreamwork, fantasy, speculation and escapism. (...) And for some tourists such visits are times and places for play. (Light 2009: 247)

Reijnders refers to these expectations of a certain place as a 'geographic imagination':

> Each of us carries an imaginary map of the world in our head, which we use to position ourselves with regard to other regions, countries, and continents. Even though we have usually not actually visited these places, we can still bring a picture of them to mind. (...) Films, news, broadcasts, TV series, comic books, games, current event programmes all help create an image of the world around us. (Reijnders 2009: 171)

These imaginary maps are created by privileging certain parts of that place above others, making some elements more characteristic of the place than others. The spaces that are featured in these narratives will from now on occupy a more privileged position on the Camino, privileged above other spaces one can encounter on the way: the albergue that was featured in a popular movie would hold a privileged position above the other three albergues down the road, and the little church that a beloved protagonist had engaged with would hold a privileged position above the dozens of other churches the pilgrim will pass. They will be identifiable markers that indicate the Camino, while other features of the route will remain without specific Camino-related significance. Reijnders points out that: 'By identifying the locations in this way (...) imaginary entryways are opened to other, diegetic worlds. In other words, the *lieux d'imagination* are being symbolically marked' (Reijnders 2011: 44). And so it happens that one familiar albergue or church will indicate more Camino-ness than the other non-familiar albergues and churches that are also placed alongside the Camino.

In many obvious ways the contours of the Camino are also literally marked. Pilgrims know from the outset to follow the yellow arrow and the St. James' scallop. These types of symbols create a sacred geography of which the pilgrim knows the markers from a very early stage. However, this process of indication goes beyond spatial points only. A similar claim could be made for themes, motifs, images and ideologies that are featured in these narratives. Themes like adventure and healing, motifs like communal dinners and the constant concern with finding a place to get coffee, images like the yellow arrow that points the way, ideologies of inclusiveness and personal growth – all of these serve as pointers that inform the pilgrim about their journey. And so the narrative that prepared the pilgrim for their journey will guide their gaze and their attention on the Camino. What are they looking for? Are they in awe of the exotic Spanish countryside, and intrigued by these funny European traditions? Or are they looking for healing in a diverse community of fellow-pilgrims?

So, when pilgrims have arrived on the Camino they start to recognize and look out for elements they have become familiar with through pilgrim narratives. However, 'recognition' is not the only relation pilgrims can have with these elements. Coleman and Elsner assert that:

> Travelers may find either what the texts have prepared them to find, or they construct a kind of anti-structure that finds the opposite of the textual range. This opposite is, of course, equally determined by the texts. (Coleman and Elsner 2003: 14)

The experiences on the Camino can also deviate from expectations. This can result in a range of different emotions: excitement, disappointment, bewilderment, relief, agitation, etc. Using examples from his own field work, Reijnders argues that these discrepancies between imagination and reality surprisingly do not undermine the authenticity of the narrative. When a traveler encounters such a discrepancy, they are confronted with the imaginary dimension of the narrative that had shaped their expectation.

> Rather than diminishing the locations, these 'bloopers' actually emphasize the authenticity of the location compared to the authenticity of the imagination. These small differences actually serve to strengthen the reciprocity between these two worlds. (Reijnders 2011: 45)

By encountering moments that emphasize the boundary between imagination and reality, pilgrims find again that they are now actually on the Camino, they are now actually pilgrims. Their pilgrimage is no longer mostly fantasy, they are actually living that fantasy. Simultaneously, pilgrims start to find out that they are becoming quite the Camino expert, they know better than the stories, and so their pilgrim identity is re-affirmed. Perhaps that is the reason that a popular pastime among pilgrims is to discuss the ways in which famous narratives were accurate or not in their portrayal of the Camino: laughing at those characters in *The Way* who could walk the Camino unprepared with someone else's shoes, who could arrive at nighttime and were still accepted in albergues – all features that do not comply with reality as it is found on the Camino.

Most experiences on the Camino are a mix of finding similarities and comparing differences. During their pilgrimage, pilgrims will come across narrative elements that were or were not a part of their pre-pilgrimage imagination. These encounters offer opportunities to personally engage with these elements and thereby to inscribe themselves into that narrative. In this engagement with mediated elements, pilgrims find themselves in an interesting place between fiction and reality: they actualize narratives that were formerly based on imagination. A media narrative, even if it strives for perfect accuracy in its representation, will always be that: a representation, and therefore lacking the reality that lies in an embodied experience. Pilgrims give a reality to these representations by embodying them, giving them life through their actual emotional engagement. So, on the one hand, the pilgrim 'plays at being pilgrim' by following the script of pilgrim narratives, and on the other hand the pilgrim extends the imagination of the narratives with their own personal presence and experience. In the actualization of these elements, all the complexities of the different themes and motifs of the pilgrim identity come into play.

Narrative Communities

The narratives that inspired people to become pilgrims can also play a role in the forming of communities along the Camino. Narratives that are particularly popular among pilgrims can create a sense of community because they allow pilgrims from different backgrounds to connect with each other. In discussions on the Camino, pilgrims often refer to narratives of other pilgrims to illustrate their point ('This reminds me of something that happened to Kerkeling on his journey – did you read that book?'). In this way, these famous narratives create common ground through which pilgrims find themselves connected to one another. The connection that is established through very well-known narratives is looser than a mutual acquaintance with very marginal narratives.

Pilgrims can also feel connected to the protagonists of the Camino narratives they read. This feeling often increases when pilgrims have started their own pilgrimage. They understand much better what the author meant when they described certain emotions or experiences, because the pilgrim finds her/himself in the same situation now. This sense of recognition and connectedness becomes even more interesting when the protagonist the pilgrim identifies with is a fictional character. For pilgrims tend not to distinguish between fictional or real characters when they discuss these feelings. When talking about their experiences of physical exhaustion, spiritual awakening, etc. they refer just as easily to characters of *The Way* as they do to pilgrims portrayed in a TV documentary, or even to other pilgrims they meet along the Camino. On my first night on the Camino Francés in 2014, during dinner at an albergue in Saint-Jean-Pied-de-Port, I talked to two pilgrims who were convinced they would be bitterly disappointed on their arrival in Santiago, because the protagonist in the novel they had both read in preparation for their journey had felt that way. Therefore, these pilgrims had made sure to have another week off from work after arriving – to have some time to come to terms with their already anticipated disillusionment inspired by the story of a fictional pilgrim. Here again, we find a mixing of imagination and reality that pilgrims don't seem to experience as problematic. These fictional characters easily become part of the Camino family.

What is more, this connection does not have to be explicitly personal; it can also be experienced in a more general kind of way. Using terminology that combines the real with the imaginary, pilgrims often talk about 'walking in the footsteps of many'. This image combines the practical reality of sharing a road with the more imaginary idea of sharing a journey:

> The route from Tours, Vezelay, and le Puy. From this place thousands of pilgrims since the Middle Ages journeyed to Jean Pied de Port. We are stepping in the footsteps of all these people from the past. (Pilgrim blog 2016)

It relates not only to the many pilgrims that surround us now, but it opens up a view of the pilgrims that have walked the way in very different circumstances – but still with similar objectives. This connection to pilgrims throughout the history of the Camino is a recurring theme and a powerful one for many pilgrims. One pilgrim writes in her blog:

> This morning I passed Molinaseca, with a beautiful old Romanic bridge. I sat there with my cup of tea looking at the bridge, and i could just SEE as it were the little feet of all those pilgrims throughout the years, centuries and centuries walking over it. All those people with their hope, their fear and happiness and sadness. So very touching I find that... (Pilgrim blog 2006)

In connecting to the medieval pilgrim, contemporary pilgrims relate to a figure that they only know through stories. The figure is not imaginary in the sense that they never existed, but they are an imagined companion on the Camino in the sense that they are not actually present. Still, their presence is felt by contemporary pilgrims. In his account of his travels through Spain, Dutch writer Cees Nooteboom remarks the following about his arrival at the cathedral in Santiago:

> It is impossible to prove and yet I believe it: there are some places in the world where one is mysteriously magnified on arrival or departure by the emotions of all those who have arrived and departed before... (Nooteboom 2000 [1992]: 1)

In conversations about the medieval pilgrim, contemporary pilgrims often stress the differences between the people traveling then and now: the danger of the journey then, the explicitly religious motivations of the medieval pilgrim, but also the luxury of credit cards and high-tech hiking gear today. This imagery is continued by popular media like guidebooks, which frequently provide descriptions of routes as follows:

> It is very evident that the road as it existed before is lost beneath a layer of asphalt or the extensive cultivation of agricultural land. Where there used to be forests, wolves, brigands, there are industrial areas and cars. And the pilgrims, who used to live off alms, now carry around their credit cards. (Nadal 2008: 8)

However, despite underlining the difference between the medieval and the contemporary pilgrim, it is generally agreed that the connection between the two figures is of higher importance, that connection being that both traveled the Camino to Santiago de Compostela.

Conclusion

The Camino de Santiago has gone through a number of changes since the grave of the apostle was found on the shores of Galicia many centuries ago. An important characteristic of its present manifestation is the freedom that is offered to pilgrims to establish the meaning of the journey – religious narratives are mixed with political, historical and leisurely narratives. In response to this freedom in interpretation, pilgrims turn to media narratives to guide them. These narratives might present pilgrims with different or unconventional interpretations of what it means to be a pilgrim. These narratives guide pilgrims along the Camino and point out what elements are significant and worthy of the pilgrim's attention. These elements might be material (an albergue, a church, a restaurant) or immaterial (friendship, healing, hospitality). By following these narrative guides, pilgrims travel the Camino twice: once as physical pilgrims traveling the Camino, and once through the diegetic worlds of the stories. They can find one another in both of these worlds and create communities through engagement with fictional characters in the narratives or actual pilgrims on the road or a combination of both. The development of the Camino de Santiago into a media pilgrimage is therefore a significant move for this centuries-old pilgrimage that could invite a new way of understanding religious or spiritual travel in the twenty-first century.

Notes

1. The research behind this chapter has been done within the context of my doctoral dissertation. Parts of this text are included in that manuscript (van der Beek 2018).
2. The prominent representation of South Koreans on the Camino is presumed to be the result of the high percentage of Christians in this country, compared to other countries in that region. According to a 2014 Pew Research report, 29 percent of South Koreans were Christian in 2010 (Connor 2014).
3. For my doctoral project, I performed several months of fieldwork on the Camino de Santiago between 2013 and 2017 (van der Beek 2018).

Bibliography

American Pilgrims on the Camino (2017), 'Compostelas issued by the Oficina de Acogida de Peregrinos, Yearly Totals 1986 through 2016', *Camino Statistics*. On the website of the American Pilgrims on the Camino. 11 February, www.americanpilgrims.org/assets/media/statistics/compostelas_by_year_8616.pdf (accessed 16 August 2019)

Beek, S. van der (2018), 'New Pilgrim Stories', PhD diss., Tilburg University.

Beeton, S. (2010), 'The Advance of Film Tourism', *Tourism and Hospitality Planning & Development*, 7 (1): 1–6.

Chemin, E. (2012), 'Producers of Meaning and the Ethics of Movement: Religion, Consumerism, and Gender on the Road to Compostela', in W. Jansen and C. Notermans (eds), *Gender, Nation and Religion in European Pilgrimage*, 37–53, Surrey: Ashgate.

Cohen, E. (1979), 'A Phenomenology of Tourist Experiences', *Sociology*, 13 (2): 179–201.

Cohen, E. (1992), 'Pilgrimage Centers: Concentric and Eccentric', *Annals of Tourism Research*, 19: 33–50.

Cohen, E. and R. L. Cooper (1986), 'Language and Tourism', *Annals of Tourism Research*, 13 (4): 533–64.

Coleman, S. and J. Elsner. (2003), *Pilgrim Voices: Narrative and Authorship in Christian Pilgrimage*, New York: Berghahn Books.

Connell, J. (2012), 'Film Tourism: Evolution, Progress and Prospects', *Tourism Management*, 33 (5): 1007–29.

Connor, P. (2014), '6 Facts about South Korea's Growing Christian Population', *Fact Tank*. On the website of the Pew Research Center. 12 August, https://www.pewresearch.org/fact-tank/2014/08/12/6-facts-about-christianity-in-south-korea/ (accessed 26 March 2020).

Costen, M. (1993), 'The Pilgrimage to Santiago de Compostela in Medieval Europe', in I. Reader and T. Walter (eds), *Pilgrimage in Popular Culture*, 137–54, London: Macmillan.

Frey, N. (1998), *Pilgrim Stories: On and off the Road to Santiago: Journeys along an Ancient Way in Modern Spain*, Berkeley: University of California Press.

Gemzoë, L. (2012), 'Big, Strong and Happy: Reimagining Femininity on the Way to Compostela', in W. Jansen and C. Notermans (eds), *Gender, Nation and Religion in European Pilgrimage*, 37–53, Surrey: Ashgate.

Herwaarden, J. van (2002), *Between Saint James and Erasmus: Studies in Late-Medieval Religious Life – Devotion and Pilgrimage in the Netherlands*, Leiden: Brill.

Instagram, 'About Us', viewed on 18 September 2017, www.instagram.com/about/us/ (accessed 16 August 2019).

Kemp, W. (2011–present), *De Camino: Op weg naar Santiago de Compostela*, [TV programme], Omroep Rooms-Katholiek Kerkgenootschap.

Kerkeling, H. (2009), *I'm Off Then: Finding and Losing Myself on the Camino de Santiago*, translated by Shelley Frisch, New York: Free Press.

Light, D. (2009), 'Performing Transylvania: Tourism, Fantasy and Play in a Liminal Place', *Tourist Studies*, 9 (3): 240–58.

Lois González, R. C. (2013), 'The Camino de Santiago and its Contemporary Renewal: Pilgrims, Tourists and Territorial Identities', *Culture and Religion*, 14 (1): 8–22.

Lopez, Lucrezia, David Santomil Mosquera and Ruben Lois González (2015), 'Film-Induced Tourism in the Way of Saint James', *AlmaTourism Journal of Tourism, Culture and Territorial Development*, 6 (4): 18–34.

Nadal, Paco (2008), *Te voet naar Santiago de Compostela*, translated by Marjan Blumer and Josine Stobbe, Delft: Uitgeverij Elmar.

Nooteboom, C. (2000) [1992], *Roads to Santiago*, translated by Ina Rilke, Amsterdam: Mariner Books.

Oficina de Acodiga al Peregrino (2017), 'La Peregrinación a Santiago en 2016', oficinadelperegrino.com/en/statistics/ (accessed 16 August 2019).

Reijnders, S. (2009), 'Watching the Detectives: Inside the Guilty Landscapes of Inspector Morse, Baantjer and Wallander', *European Journal of Communication*, 24: 165–81.

Reijnders, S. (2011), *Places of the Imagination: Media, Tourism, Culture*, Farnham: Ashgate.

Riley, R. and C. S. van Doren (1992), 'Movies as Tourism Promotion: A "Pull" Factor in a "Push" Location', *Tourism Management*, 13 (3): 267–74.

Talbot, L. (2015), 'Revival of the Medieval Past: Francisco Franco and the Camino de Santiago', in Sánchez y Sánchez and Annie Hesp (eds), *The Camino de Santiago in the 21st Century*, 36–56, New York: Routledge.

Turner, V. (1974), *Dramas, Fields, and Metaphors: Symbolic Action in Human Society*, Ithaca: Cornell University Press.

Turner, V. and E. Turner (1978), *Image and Pilgrimage in Christian Culture: Anthropological Perspectives*, New York: Columbia University Press.

Voragine, J. de (1969), *The Golden Legend*, translated by R. Granger and H. Ripperger, New York: Arno Press.

Chapter 7

Cultural-religious Routes and Their Tourism Valorization: 'In the Footsteps of the Apostle Paul' in Greece

Polyxeni Moira

1. Introduction: Cultural Routes

Apostle Paul, a Greek-speaking Jew who converted to Christianity during the second half of his life, is regarded as one of the most influential people in the history of Christianity, as out of the twenty-seven books of the New Testament, which is the second part of the Scripture, thirteen are attributed to Paul. A large part of his missionary travels, which included preaching to the Gentiles, took place in the first century CE in Ancient Greece, which was then dominated by the Roman Empire. Greece has emerged as one of the world's top tourist destinations, exceeding 33 million international tourist arrivals in 2018 (UNWTO 2019) and boasting eighteen UNESCO World Heritage Sites. This chapter explores how Eastern Orthodox Christian practices in Greece have been integrated into the Council of Europe's cultural route, named 'In the Footsteps of Apostle Paul', which displays the nexus between cultural-religious heritage, tourism and sustainable development. The European Cultural Route 'In the footsteps of St Paul, the Apostle of the nations', was formally launched by the Cult-RInG Project lead partner, as a candidate cultural route for certification by the Council of Europe, on 24 April 2020 (CoE: 2020).

Cultural Routes are one of the most important tools for the promotion and interpretation of cultural heritage. They are itineraries designed around a core theme and they include, inter alia, monuments of archeological or historical interest, architectural monuments, industrial heritage buildings, religious heritage edifices and traditional settlements. The connecting element or the shared theme might be the category of monuments (e.g. religious pilgrimages, cemeteries, castles etc.), the time period these monuments are associated with (e.g. antiquity, Middle Ages) or their geographical scope. An interdisciplinary approach is essential for the design and conceptualization of Cultural Routes, which means that the theme must be researched and developed by multidisciplinary experts and the major single sites and points of interest identified. Then techniques concerning the necessary intervention at the monuments, sites and surrounding areas are designed and implemented. Finally, cultural initiatives and projects as well as tourist products are developed for the sound management, operation and promotion of the cultural routes.

Wided Majdoub (2010: 30) identifies cultural routes and itineraries as 'a relatively recent cultural phenomenon that led to the emergence of a new type of heritage', and defines them as 'both a geographical journey through a territory and therefore through plural local identities, but also a mental journey with representative values, meanings, expectations, experiences, and finally a tourism product'.

In 1987 the Council of Europe launched the Cultural Routes program, with the objective of demonstrating, by means of a journey through space and time, how heritage of different European countries and cultures contributes to a shared cultural heritage. Moreover, the fundamental principles of the Council of Europe such as human rights, cultural democracy, cultural diversity and identity, dialogue, mutual exchange, and enrichment across boundaries and centuries are promoted (CoE 2010, 2017).

The definition of cultural routes is given by the International Scientific Committee on Cultural Routes (Comité Internacional de Itinerarios Culturales/CIIC) of the International Council on Monuments and Sites (ICOMOS) in the ICOMOS Charter on Cultural Routes,

> Any route of communication (…) must fulfill the following conditions: a) It must arise from and reflect interactive movements of people as well as multi-dimensional, continuous, and reciprocal exchanges of goods, ideas, knowledge and values (…), b) It must have thereby promoted a cross-fertilization of the affected cultures in space and time (…) and c) it must have integrated into a dynamic system the historic relations and cultural properties associated with its existence. (2008 [2003]: 3)

All the elements of the sites and areas along a cultural route jointly compose a complex system in a succession of different landscapes (religious, industrial, rural, urban and maritime), which helps in the interpretation and the understanding of the diversity of the identities which make up the continent of Europe. At the same time nowadays, cultural routes are brought forward as a significant tourism product and by extension as an engine for economic and regional development. However, as mass tourism continuously grows and evolves, there is a hyper-concentration of tourists in particular sites and areas where specific heritage sites are located. This situation risks misinterpretation of cultural goods and their contents. In other words, cultural routes have the ability to function as complex and dynamic systems and tools for contextualized interpretations of the heritage sites. Cultural routes are meant to attribute cultural goods with multiple layers of content and establish relationships among them, which allows a larger understanding through thematic continuities. Thus, cultural routes are regarded as a necessary tool for substantial interpretations of a wider and more sophisticated cultural heritage at the regional level.

Marina Karavasili and Emmanuel Mikelakis claim that '[…] the cultural route is one of the most widespread management tools for the development of cultural tourism […] and suggest a predefined trail (visit) to monuments of natural and cultural heritage, which are integrated into a joint thematic historical or conceptual framework. Cultural routes are therefore an applied practice for the interpretation of cultural heritage' (1999: 50).

2. The Religious Cultural Routes

On 23 October 1987, the Santiago de Compostela Declaration was signed and the Santiago De Compostela Pilgrim Routes were recognized as a European Cultural Route, making it the first labeled with that designation. The Camino de Santiago (also known as the Way of St. James) was the first cultural route that served as a reference point for other European cultural routes designed and implemented subsequently (Fisher 1992: 76). The Cultural Routes represent journeys through time and space which help us keep sight of where we came from, and who we are (Battaini-Dragoni 2015).

Currently, the program includes thirty-one certified cultural routes and extends throughout continental Europe. The number of grassroots networks and associations, local and regional authorities, universities, and professional organizations involved is continuously increasing. The harmonious coexistence of tourism and culture, coupled with the possibility of exploring cultural heritage, is considered a strong trend in tourism, since it is about a new approach based on local cultural values related with an aspect of the diverse European identity. This program is the core of a tourism product whose success is getting bigger and bigger. Moreover, every year hundreds of events and educational exchanges take place along the routes and local projects are implemented, with the participation of thousands of citizens from 900 local communities.

In addition to the Council of Europe, which certifies cultural routes through the European Institute of Cultural Routes (EICR),[1] UNESCO has included two Religious-Cultural routes on the World Heritage List: the Routes of Santiago de Compostela (*Camino Francés* and Routes of Northern Spain) in 1993 and the Sacred Sites and Pilgrimage Routes in the Kii Mountain Range in Japan in 2004.

Some cultural routes are closely associated with the life, works and death of religious-spiritual personalities who preached their faith. Thus, travelers are encouraged to make a religious-spiritual-cultural journey, by following the trail of a martyr, a prophet or a historic person who sought the divine presence. At the same time travelers explore monuments and sites dedicated to this personality, recalling the historical, social and religious-cultural context in which they lived (Moira 2020: 248).

The religious-cultural routes are very diverse. It does not only concern the life of a saint or a martyr but also architecture, art, agricultural practices, techniques and methods of harnessing the natural environment in order to adapt it to human needs. Therefore, a religious-cultural route may be related to faith (e.g. Santiago de Compostela Pilgrim Routes), memorialization (e.g. the European Cemeteries Route)[2] and information on history, art, architecture, technical works and lifestyle (e.g. the Romanesque Routes, the European Route of Cistercian abbeys) or a combination of these.

Pilgrims' motivations to undertake the Camino de Santiago have been shown to vary from religious (e.g. by faith, participating in sacraments) to spiritual (e.g. enjoy solitude and inner peace, experience a more simple lifestyle) and cultural (e.g. visit historical places, learn the culture of other places) motivations (Antunes, Amaro and Henriques 2017).

Religious-cultural routes reflect the diversity of the cultural and spiritual identities of the peoples and represent a fertile breeding ground for an inter-cultural and inter-religious dialogue.

The creation of cultural routes can be based on three general criteria: the historical, geographical and thematic. In accordance with the historical criterion, a cultural route may cover: (a) The history and the culture of a specific time period (e.g. the Prehistoric Rock Art Trails Association, certified in 2010 as a Cultural Route of the Council of Europe); (b) The course of a historic person (e.g. Via Francigena) (CoE 1994); (c) The monuments and the areas which relate to the life, cult or folklore of important personalities (e.g. the Saint Martin of Tours Route) (CoE 2005); or (d) A geographical area specialized in the production of a specific product (e.g. the Pyrenean Iron Route as the Pyrenees region is rich in iron ore). In accordance with the geographical criterion, the cultural route may be: (a) *Transnational*, including tours in sites, museums and fortresses across several countries (e.g. the Route of the fortified towns of the Grande region [2016] between France, Germany, Belgium and Luxembourg featuring European military architecture); (b) *Regional*, covering monuments of cultural tradition spread over a wider geographical area (e.g. the Viking Routes); or (c) *Mediterranean*, including visits to sites connected with the production of a common product among the Mediterranean peoples (e.g. the Routes of the Olive Tree).[3] And, finally, in accordance with the thematic criterion, a cultural route may refer to: (a) *Religious/pilgrimage cultural routes* (e.g. the Santiago de Compostela Pilgrim Routes); (b) *Commercial routes* which may specialize in the transport of specific products (e.g. Silk Roads); or (c) *Historic routes* of territorial conquests (e.g. Via Habsburg), trade exchanges (e.g. the Phoenicians' Route) or historic persons (e.g. Destination Napoleon, the European Routes of emperor Charles V). To illustrate these different criteria, Table 7.1 lists all religious-cultural routes of the Council of Europe.

Table 7.1. The cultural-religious Routes of the European Institute of Cultural Routes (EICR). Based on CoE 2018b, processed by author.

Cultural Route	Theme	Year of Launch	Type of Route	Participating Countries
The Santiago de Compostela Pilgrim Routes	The legend holds that St. James's remains were carried by boat from Jerusalem to northern Spain, where he was buried on what is now the city of Santiago de Compostela.	1987	Religious	Belgium, France, Italy, Spain

Cultural Route	Theme	Year of Launch	Type of Route	Participating Countries
The Via Francigena	In AD 990, Sigeric, Archbishop of Canterbury, traveled to Rome to meet Pope John XV and receive the investiture pallium. Along the way, he recorded the 79 stages of the journey in his diary. Thanks to this document, it has been possible to reconstitute the then shortest route between Canterbury and Rome, which can now be followed by all travelers.	1994	Historical or religious	France, Italy, Switzerland, United Kingdom
The Cluniac Sites in Europe	In the early 10th century, William the Pious, Duke of Aquitaine, founded a Benedictine Abbey in Cluny, in the French region of Burgundy. During the Middle Ages, Cluny became a major center of European civilization, resulting in the emergence and development of over 1800 sites throughout western Europe.	2005	Religious	France, Germany, Italy, Spain, Switzerland, United Kingdom
The Saint Martin of Tours Routes	Saint Martin of Tours is one of the most familiar and recognizable Christian saints and has been venerated since the fourth century. He was the Bishop of Tours, whose shrine in Gaul/France was the target of a very important pilgrimage, the equivalent of that to Rome, during the Early Middle Ages, before becoming a famous stopping-point for pilgrims on the way to Compostela.	2005	Historical or religious	Austria, Belgium, Croatia, France, Germany, Hungary, Italy, Luxembourg, Serbia, Slovenia, Spain

Cultural Route	Theme	Year of Launch	Type of Route	Participating Countries
St. Michael's Way	The route was created after the discovery of ancient shipping rosters proving that pilgrims historically came to Lelant by sea on their way to Santiago de Compostela and traveled over land to Marazion to avoid the waters of Lands' End.	1994	Religious	United Kingdom
Transromanica -The Romanesque Routes	Around the year 1000, artists from all over Europe were inspired by the Roman and early Christian tradition, giving birth to a unique architectural style: the Romanesque. The Romanesque style incorporated local myths and legends to reinvent old traditions, thus reflecting the specific geographic characteristics of each region of medieval Europe over a period of 300 years.	2007	Historical or religious	Austria, France, Germany, Italy, Serbia, Slovenia, Spain
The European Cemeteries Route	Cemeteries are part of our tangible heritage, for their works, sculptures, engravings, and even for their urban planning. Cemeteries are also part of our intangible heritage, our anthropological reality, providing a framework surrounding the habits and practices related to death.	2010	Historical or religious	Austria, Croatia, Estonia, France, Germany, Greece, Italy, Norway, Poland, Portugal, Russian Federation, Serbia, Slovenia, Spain, United Kingdom

Cultural Route	Theme	Year of Launch	Type of Route	Participating Countries
The Route of Saint Olaf Ways	Olav II Haraldsson, later known as St. Olav, was King of Norway from 1015 to 1028. After he fell in the battle of Stiklestad in 1030 he was declared a martyr and a saint, which led to the propagation of his myth. For centuries after his death, pilgrims made their way through Scandinavia, along routes leading to Nidaros Cathedral, in Trondheim, where Saint Olav lies buried.	2010	Historical or religious	Denmark, Norway, Sweden
The European Route of Cistercian abbeys	The Cistercian Order represents a rich legacy that is still present today at the heart of the Roman Church and European states. The "white monks" were and still are exemplary constructors, participating in the development of rural areas by controlling the most advanced hydraulic and agricultural techniques - through their barns, cellars, mills and foundries - and have contributed to the development of art, knowledge and understanding in Europe since the Middle Ages.	2010	Historical or religious	Belgium, Czech Republic, Denmark, France, Germany, Italy, Poland, Portugal, Spain, Switzerland, Sweden

Cultural Route	Theme	Year of Launch	Type of Route	Participating Countries
The European Route of Jewish Heritage	The European Jewish heritage is widely present across Europe. Notable examples include archaeological sites, historic synagogues and cemeteries, ritual baths, Jewish quarters, monuments and memorials. In addition, several archives and libraries, as well as specialized museums devoted to the study of Jewish life, are included in the route. This route fosters understanding and appreciation of religious and daily artefacts and also recognition of the essential role played by the Jewish people in European History.	2004	Historical or religious	Belgium, Croatia, Czech Republic, France, Hungary, Italy, Lithuania, Luxembourg, Netherlands, Serbia, Slovakia, Spain, Ukraine, United Kingdom

In Greece, apart from the existing cultural routes certified by the Council of Europe, there are also national religious-cultural routes, such as the 'Routes of the Orthodox Monasteries'[4] and the route, central to this chapter, that follows in the footsteps of the Apostle Paul in Greece.

3. Saint Paul in Greece

Saint Paul's missionary journeys offer a relevant case study of contemporary Bible-related tourism. I will first describe the places in Greece that Paul traveled to and preached in. This provides a background for understanding the Cultural Route created on this theme.

In the middle of the first century CE, an important change occurred to the Greco-Roman world, with the alliance of Greek spirit and emerging Christianity. This alliance was crucial for the evolution of European civilization – and beyond. Over this period Roman culture was greatly influenced by Greek culture throughout the Roman Empire, through language, education, arts, institutions and social values. At the same time early Christianity through the Greek language and education introduced innovative ideas about God, the world and man. People were taught and inspired by the messages of the Gospels. Jesus' teaching had an impact on people's lives, societies and culture, as it allowed uncovering the importance of the human being. In Christianity, the human personality was recognized and became respected. Paul was the first to combine Greek culture and Christianity in his teachings, laying the foundations of a new world discourse. He was called the Apostle of the Nations, as he received the mission of going and proclaiming the word of God to all people in Palestine, Greece, Cyprus, Italy, including diaspora Jews and non-Jews (Holy Metropolis of Thessaloniki, n.d.: 7).

Saint Paul (Saul of Tarsus of Cilicia, ca. 5 CE–ca. 67 CE) was not a disciple of Jesus, one of the Twelve Apostles, but he is generally considered one of the most important figures of the Apostolic Age. He began teaching the gospel of Christ sixteen years after the death of Jesus, and over the next twenty years he made four missionary journeys in the Mediterranean, the Middle East and Europe, according to the accounts of Luke, the author of the Gospel and Acts, and the Epistles of Saint Paul. There are also records of a fourth, non-missionary journey to Rome. After spending roughly two years in Caesarea's prison, Paul requests, in 60 CE, that Roman Governor Festus send him to Rome to have the case against him heard by Caesar (Despotis 2009: 321).

During his missionary journeys, Saint Paul only visited large cities to preach the Holy Gospel, and to found a local *Ekklesia*.[5] He traveled along the main trade roads by land or by sea, and the three major centers of his missionary activity were Antioch on the Orontes, Ephesus and Corinth. Travelling to large commercial centers and provincial capitals permitted Paul to address large audiences.

Paul visited Greece on his second and third missionary journeys, although some claim that he visited Greek cities on his fourth journey as well. Some scholars regard this as the most prolific stage of his life, during which Christianity was first introduced into Europe where it has been practiced ever since (Despotis 2011: 11).

The second missionary journey of the Apostle Paul took place under Claudius between 49 CE and 52 CE (Acts 16; 17; 18). This journey marks the presence of Christian preaching in Europe for the first time. Paul revisited the *Ekklesia* he had established in Asia Minor on his first journey and also founded *Ekklesia* in Greece. He set out from Antioch-on-Orontes, arriving in Derbe and passing through Lystra, Phrygia, Galatia[6] and Troas, and from there traveling to the island of Samothrace in the Aegean Sea. Then he traveled to Neapolis, Philippi, Thessalonica, Berea, Athens and Corinth, before completing his return journey to Jerusalem via Ephesus and Caesarea.

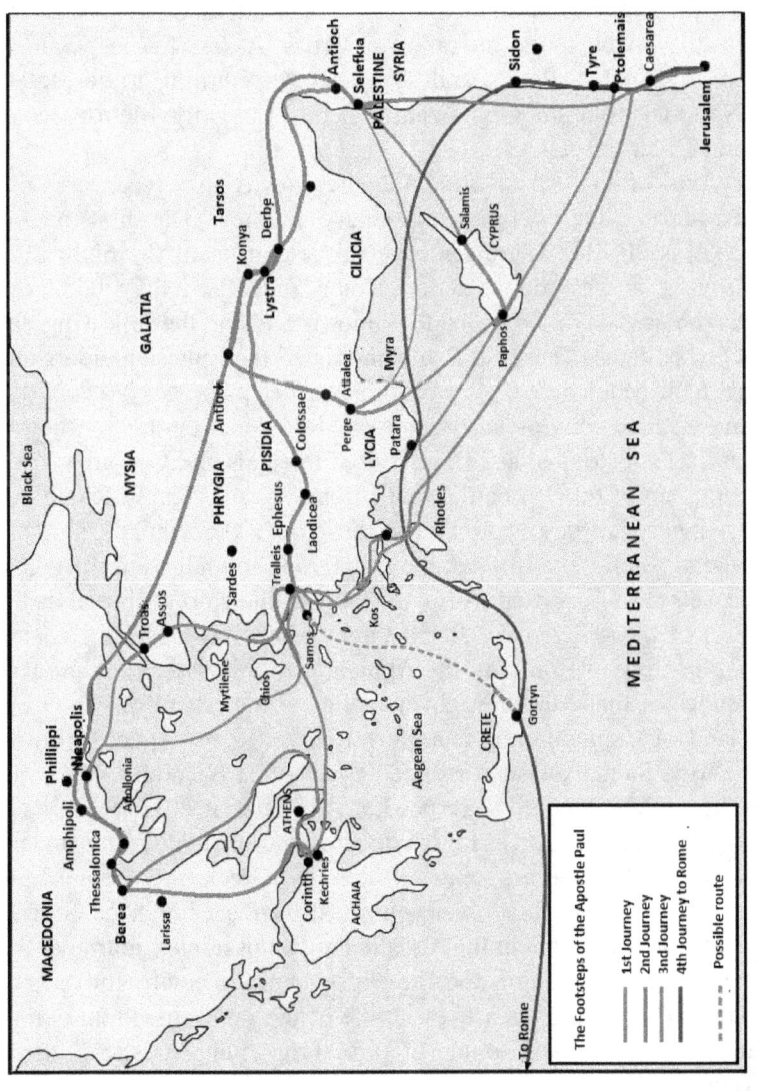

Figure 7.1. The Footsteps of the Apostle Paul (Missionary journeys).
Source: Geraki 2006. Translated from Greek to English by the author.

The third missionary journey of Saint Paul took place between 52 CE and 56 CE (Acts 19; 20). After he stayed in Antioch (Syria) for some time, Paul set out for Phrygia and Galatia, with a view to supporting the *Ekklesia* he had founded there on his second journey before arriving in Ephesus. From there he traveled to Troas and then visited Macedonia (Neapolis, Philippi, Thessalonica, Berea), Corinth and went back to Macedonia. From Neapolis he sailed to Troas, Assos, Lesbos, Samos, Miletus, Cos, Rhodes, Patara and Tyre, before returning to Jerusalem (Greek National Tourism Organization 2003: 12; Holy Metropolis of Thessaloniki, n.d.: 20–1).

In the Acts of the Apostles some other towns in Greece are also mentioned during Paul's third missionary journey. The first one is Mytilene (Acts 20.14), where Paul and his escort spent the night after departing Assos and on their way to Chios island, where they also spent the night. The next day they sailed to Samos island and the following day they reached Miletus. There St. Paul summoned the Ephesian elders and delivered on the beach one of his most impassioned messages (Acts 20.17-18). Then he sailed for Cos island and Rhodes island. On his journey to Jerusalem St. Paul stopped at Patara, Tyre, Ptolemais and Caesarea.

In the autumn of 60 CE, Paul set sail from Myra, Lycia, to Rome, his fourth missionary journey. Nevertheless, because of the wind, the ship did not manage to reach Italy. Instead, the ship turned south toward the island of Crete (Acts 27.3-7). After a difficult journey the ship anchored at the Cretan port of Fair Havens.[7] After a three-month stay in Meliti island, Paul arrived safely in Rome by the Appian Way (Via Appia). A modern theory advocates that Melite might not refer to the island of Malta but to the island of Cephalonia (Metallinos 1999). Moreover, according to another theory, Saint Paul also stopped by at Actia Nicopolis of Epirus Vetus in the Ambracian Gulf (Tzerpos n.d.: 7; Meletios 2015).

The most referenced sources for the documentation of Paul's missionary journeys within Greece are texts written by Greek theologians and university tutors, which have been approved by the Church of Greece. The first reliable source about the Apostle Paul's missionary journeys and the content of his preaching is the New Testament, especially the Acts of the Apostles. This text gives a first outline of the rapid dissemination of Christianity throughout the Roman Empire, through the Apostles' actions (Goutzioudis n.d.: 6).

Several studies analyze the different stopovers during Paul's missionary journeys. For example, in *Apostle Paul in Athens* (2009) Sotirios Despotis deals with Paul's journey to Athens and his speech on Arios Pagos.

Despotis seeks to educate readers about the prevailing conditions of the time and the philosophical currents that St. Paul encountered there in 50 CE. Paul, seeking to adapt his teaching to the Athenian audience he addressed, added to his Areopagus sermon quotes from the Old Testament coupled with contemporary meaningful elements. In addition to the twelve main gods, ancient Greeks worshipped a deity they called 'Unknown God'. Paul tried to deter the Athenians from worshiping Unknown God and encouraged them to recognize him and worship him as the Christian God

This relativist and dialogic philosophy of Paul's sermon seems to resonate well with modern notions of interreligious and intercultural contacts and identities.

Others refer to Paul's Epistles which seek to examine his preaching on the Anthropos and highlight the ontological criterion of understanding and interpretation of the Anthropos, or else make a comparative study of his preaching about the world within the context of the worldview problems of ancient Greek philosophy (Ioannides 2009; Oikonomou 2008).

4. The Tourism Valorization of St. Paul's Religious-Cultural Heritage

Religious-cultural heritage has always been highly valued by the Greek state. Conservation and protection of tangible and intangible religious-cultural assets and resources are regarded as essential in the promotion of national memory and identity. Nowadays, the Greek state integrates religious-cultural heritage into the tourist sector, aiming to valorize this heritage by trying to diversify the tourism product on the one hand and contribute to an understanding of Europe's cultural diversity on the other. Valorization means the creation of information channels to inform the present through the interpretation of the past, taking into account the particular relationship that has existed in Greece between the church and state until today. Deno John Geanakoplos highlights that '[i]n the medieval theocratic societies of both the Byzantine East and the Latin West, where the influence of Christian precepts so strongly pervaded all aspects of life, it was inevitable that the institutions of church and state [...] be closely tied to one another' (1965: 381). Moreover, during the period of Ottoman rule in Greece, lasting from the mid-fifteenth century until the successful Greek War of Independence that broke out in 1821 and the proclamation of the First Hellenic Republic in 1822, the role of the Ecumenical Patriarchate of Constantinople, the symbolic headquarters of the Greek Orthodox Church, was vital. Dimitrios Mylonopoulos emphasizes that this undoubted interdependence between religion and state is validated in Article 3, par. 1 of the current Constitution of Greece (1978), which stipulates:

> The prevailing religion in Greece is that of the Eastern Orthodox Church of Christ. The Orthodox Church of Greece, acknowledging our Lord Jesus Christ as its head, is inseparably united in doctrine with the Great Church of Christ in Constantinople and with every other Church of Christ of the same doctrine, observing unwaveringly, as they do, the holy apostolic and synodal canons and sacred traditions. […]. (2001: 33)

Years after Paul's missionary travels, Greek Christians had churches, monasteries and monuments built in places that he toured. For example, in Neapolis (modern city of Kavala), the church of St. Nicholas ('Agios Nikolaos') was built on a Christian basilica, the remnants of which are still preserved on the path of the Apostle Paul. At the point where he first arrived in the city now stands a monument to commemorate this event and honor his first steps in Kavala. The monument covers 35 square meters and is situated in the yard of St. Nicholas church (Fig. 7.2).[8] From Neapolis Saint Paul walked along the ancient Via Egnatia, which survives to this day, and then made the twelve-kilometer journey to the city of Philippi, where he founded the first church in Europe. Other monuments include the Pilgrimage of Agia Lydia of Philippians (Baptistery of Lydia), built in 1974 in Krinides, near Kavala. Lydia, a merchant of purple dye for textiles, from Thyatira in Asia Minor, was the first woman in Europe to be baptized at the banks of the River Zygaktis outside the church. Other churches and monuments erected in his honor are St. Paul's Cathedral in Kavala, inaugurated in 1926, with a mosaic by the entrance; the memorial titled 'the Bema' in Berea (Veria), with two modern mosaics located at the top of St. Paul's Steps where it is believed that Paul addressed the crowds in 51 CE; and St. Paul's Church in the small rocky bay of Agios Pavlos on Rhodes island, where Paul is said to have stopped around 58 CE to preach Christianity. The places that he visited are not only of spiritual and religious importance, but also significant cultural monuments. Many of them are included on UNESCO's World Heritage List – e.g. the Archaeological Site of Philippi (UNESCO 2016)[9] – or they appear on the Permanent Catalogue of Listed Archaeological Sites and Monuments in Greece – e.g. Saint Paul's church in Mikros Yialos, Rhodes island[10] and the Paleochristian Basilica of the port on Kos island.[11]

Figure 7.2. The monument commemorating the event of St. Paul's disembarking in Neapolis was erected in 2000 in the courtyard of the Holy Church of Agios Nikolaos in the modern city of Kavala.
Photo by the author.

Nowadays the monuments, the churches and the sites associated with the life and preaching of Saint Paul attract thousands of pilgrims and visitors with various motivations, which fall under a variety of categories.

Pilgrimage. The churches, the monuments and the sites connected with St. Paul are part of trails followed by Christian pilgrims of all denominations (Orthodox, Catholics, Protestants, Anglicans) as well as pilgrims of other religions, who travel to Greece every year from Europe, the Americas, Africa and Asia (I Kathimerini 2017). For example, the Nigerian Christian Pilgrim Commission has organized pilgrimage-tours in the footsteps of the Apostle Paul in Greece (Athina 9,84, 2017).

Religious-Cultural Reasons. Travelers also follow the footsteps of St. Paul the Apostle because they are interested in history and culture. In this way, they combine pilgrimage with a cultural itinerary to some of the historical and most beautiful places in Greece (e.g. Athens, Corinth, Patmos island).

Religious-Cultural Events/Celebrations. These take place in the Greek cities where Paul preached the Gospel during his missionary journeys. For example, Pavlia celebrations take place in the cities of Veria and Corinth,[12] places which were central to Paul's ministry and attract a wide audience, including pilgrims. These events may include celebrating and serving at mass, prayers, litanies and vespers, venerating the relics of saints, participating in the procession of the holy icons, classical music concerts and other cultural events.

Wedding Ceremonies. Many Greek and foreign couples choose to hold, especially in the summer, their wedding ceremonies in the chapel of Agios Pavlos on Agios Pavlos beach, near the ancient acropolis of Lindos on Rhodes island, where Paul is said to have sought shelter (Proto Thema 2012). Couples, not only those getting married for the first time, but also those getting remarried or renewing their vows, are attracted by both the natural beauty and the setting of such sites, as well as the prolonged sunshine for much of the year (Major, McLeay and Waine 2010: 252). In addition to this, the setting of the wedding ceremonies provides couples with more emotional engagement during their honeymoon (Del Chiappa and Fortezza 2015: 64). In any case the tourism promotion of these destinations attracts additional visitors, pilgrims or tourists.

Participation in Sports Events – for example, the twelve-kilometer running competition entitled 'In the Footsteps of the Apostle Paul', which commemorates the itinerary from Cenchreae, the eastern harbor of ancient Corinth where Paul disembarked, to ancient Corinth. This race is held annually (now in its twentieth year) some days before the Feast of Saints Peter and Paul, which is observed on 29 June. Another twenty-kilometer track race entitled 'In the Footsteps of the Apostle Paul' is held on 20 May, the Feast Day of St. Lydia of Philippians. Runners start from the Church of St. Paul in Kavala and finish at the ancient theatre of Philippi, where Paul preached his first evangelical sermon and baptized the first Christians on European soil.[13] The route includes the visible part of the Via Egnatia and leads to (a) the archeological site of Philippi, a UNESCO World Heritage Monument (UNESCO 2016), and (b) the mud baths and thermal center of Krinides, within walking distance from Philippi.

These sporting events are co-organized by the local Holy Metropolis and the local sport and cultural association. Participants are not professional athletes but ordinary people and faithful Christians. Actually, in many instances these people do not run but walk, thus making it a sort of pilgrimage, with many participants thinking that they finish the race with the help of God. It is indicative that the medal for the race, *The passage of Saint Paul*, bears his image. The race is combined with the people's participation in religious ceremonies and a pilgrimage.

Religious Conferences. In June 1995 the 'Pavlia' five-day celebrations were launched for the first time, on the initiative of the Very Reverend Metropolitan of Veria, Naoussa and Campania, Panteleimon, to commemorate the life and work of Saint Paul. On the last three days of the annual Pavlia celebrations, an international scientific conference also takes place at different venues related with Saint Paul. For instance, in September 2007 an international conference took place in Corinth, entitled 'The Apostle Paul and Corinth': 1950 years after the writing of the Epistles to the Corinthians (Belezos 2009). There, ninety-six academic papers in English, French, German and Greek were presented by renowned scholars of Paul from across the globe. The 22nd International Scientific Conference entitled 'Apostle Paul and the Philosophers' was held in June 2017 in the context of the 22nd Pavlia celebrations in Veria, Greece (Holy Metropolis of Veria, Naoussa and Campania 2017).

5. Developing Cultural Routes

While developing a cultural route, various factors, tangible and intangible resources, and different agencies and bodies need to be taken into consideration. Because of the complicated nature of a cultural route, interdisciplinary cooperation is required (i.e. among historians, archaeologists, jurists, sociologists, geographers, civil engineers, spatial planners, environmentalists, etc.).

More specifically, the procedure followed in order to develop a cultural route includes defining the theme and the objectives of the specific route, selecting the most important monuments of cultural heritage along the route, creating a brand name, designing a logo and identifying the most appropriate ways to communicate this new product. This presupposes a knowledge of the target market, the integration of the route into the tourism planning of the area and the regular evaluation of the cultural route. In this way, the qualitative features of the route are ensured, relating to '[…] its studious spread over space and time, the protection of the natural and man-made environment (carrying and tourism capacity), [and] the safeguarding of the valorization of the cultural heritage' of each area (Karavasili and Mikelakis 1999: 51).

Another fundamental precondition for the success of this venture is to identify possible sources of funding and local and regional bodies that will undertake the promotion of the cultural route.[14] The next step is to signpost the route and design a map of the route. Additionally, road network and visitor infrastructure are essential.

In terms of promotion of a cultural route, advertising and PR are necessary. These could include designing and printing multilingual leaflets and brochures, designing a dedicated webpage for the route, uploading promotional material to social media platforms, raising awareness among the public through commercial spots in local and national media and participating in relevant international fairs and exhibitions.

Finally, a management agency has to be established. This agency will supervise the smooth operation of the route as a thematic tourism product, evaluate and monitor the aforementioned procedures, coordinate and ensure the cooperation between stakeholders without altering the sustainability of the project and guarantee the fulfillment and the success of the initial goals (Hadjinikolaou, Zirinis and Sofikitou 2015; UNWTO 2015: 60–1).

The philosophy behind the planning is that through the development of a thematic route, the concerned area turns into an attraction for a great number of visitors. This ensures the protection and conservation of the route as well as local and regional development, because of the revenue generated and the jobs created. For this reason, cultural routes are inextricably linked with tourism, local production and development.

6. Development and Certification of a Cultural Route

A religious-cultural route is designed on the tracks of ancient routes, thus restoring the religious-cultural dimension of travel and the fine connection between the nature and the spirit of those who experience it. These routes have a significant impact on the tourism industry and other economic activities as they result in the systematic organization of all the natural, cultural and financial resources available in each region. For instance, synergies and complementarities between the supplied cultural and tourism services are created (e.g. culture, tourism, agritourism, gastronomy). Moreover, the development of cultural routes contributes to the creation of networks between rural communities, thus encouraging the tourism promotion and tourism valorization of their resources.

A lot of public and private bodies, businesses, and professionals in the field of tourism and culture, at a local, national or international level are involved in the development of a cultural-religious route (Fig. 7.3). In Greece, Public bodies include the Ministry of Tourism, the Greek National Tourism Organization (GNTO), the Ministry of Culture and

Sports, the Church of Greece through the Special Synodical Committee of Pilgrimage Tours (Religious Tourism) and the Synodical Office of Pilgrimage Tours, the Holy Metropolises,[15] the Ministry of Environment and Energy, the Ministry of Infrastructure and Transport and the Local Government (Regions and Municipalities with cultural assets).

The Ministry of Culture and Sports is the government department entrusted with the preservation of the country's cultural heritage and the arts. It deals with issues related to archeological sites and monuments and provides scientific support for the development of routes in historic sites. The Archaeological Receipts Fund is a legal entity of public law under the supervision of the Ministry of Culture and Sports, by virtue of Law no. 736/1977. It is responsible for managing, marketing and promoting cultural goods. It receives, manages and allocates its funds, inter alia, for the protection, conservation, physical arrangement and valorization of archaeological sites, monuments, museums and their surrounding space.

The Ministry of Tourism grants licenses to tourism businesses, and GNTO is responsible for the promotion of Greece and its tourism product at a national level. The Ministry of Environment and Energy is responsible for issues related to protected areas, national parks, mountain routes and national trails. The Ministry of Infrastructure and Transport addresses matters pertaining to the management of the existing infrastructure and the construction of new infrastructure (road network, signposting, ports, airports) where necessary. The above departments have an institutional-regulatory role. The Synodical Office for Pilgrimage Tours of the Holy Synod of the Church of Greece promotes in Greece and abroad pilgrim tours and the development of all types of religious tourism in Greece, and valorizes the religious monuments and monasteries of the Church of Greece.

The Greek Regions are responsible for handling administrative issues along large parts of the cultural route, while the Greek Municipalities are only involved in local issues, and thus specify locally the regional cultural policy.

Moreover, the private sector is also involved in the development of a cultural route, such as local businessmen (e.g. tourist accommodation enterprises, travel agencies, motor coach companies, tour guides), tour operators, airlines (national and international) and the local community. Very often Non-Governmental Organizations (NGOs) are also involved in the protection and valorization of cultural heritage. Bodies such as the Hellenic Chamber of Hotels and the Greek Tourism Confederation (SETE) undertake to study, manage and promote the route.

If a route is awarded international certification, the role of international organizations (e.g. European Institute of Cultural Routes of the Council of Europe, UNESCO, European Network of Cultural Tourism) is significant.

Figure 7.3. Stakeholders in the recognition and award of the certification of a religious-cultural route.

On the Greek side, the Region of Central Macedonia has undertaken the initiative of applying for the award of certification of the cultural route 'In the Footsteps of St. Paul, the Apostle of the Nations – Cultural Route' as an official Cultural Route of the Council of Europe. To this end, a Memorandum of Understanding (MoU) has been signed by the Region of Central Macedonia, the Regione Lazio, Italy, the Council of Tourism of Paphos, Cyprus and the European Cultural Tourism Network-ECTN. The MoU provides for the establishment of a body with legal status, whose objective is to coordinate the actions of the stakeholders at a local and regional level (e.g. universities, regions, municipalities, Holy Metropolises, public and private agencies). This will be funded within the framework of the cooperation program INTERREG V-A Greece – Cyprus 2014 – 2020, Re-CULT project: *Valorization and dissemination of the cultural and natural heritage through the development and institutional reinforcement of Religious Tourism in Greek insular areas and in Cyprus*. An initiative for a route can be made in accordance with certain conditions: defining a theme representative of European values[16] and common to at least three European countries; identifying heritage elements shaped by the geographical as well as cultural, historical and natural features of the different regions; creating a European network with legal status bringing together the sites and the stakeholders that are part of the Route; coordinating common actions to encourage different kinds of cultural co-operation; and creating common visibility to allow the identification of the various elements of the Route, ensuring recognizability and coherence across Europe (CoE 2018a).

To be awarded certification as a Cultural Route of the Council of Europe, a complete dossier must be submitted fulfilling the criteria described in the Resolution CM/Res(2013)67, adopted by the Committee of Ministers on 18 December 2013 (CoE 2013).

The award of the Cultural Route of the Council of Europe certification is a rigorous, time-consuming and multi-level process, which is implemented through the application of objective criteria established by the Council of Europe, with the participation of experts from different disciplines. An in-depth and detailed documentation of the components of the route is required, which consists of a thorough study of the tangible and intangible elements that compose it and which furnish proof of its authenticity. A two-year certification cycle is applied for the development of a Cultural Route, which consists of following certain steps necessary for the completion of the procedure. These steps include, for example, the creation of a European network with legal status of at least three European countries, the coordination of common activities in the main field of activities and the creation of common visibility to ensure the recognizability and coherence of the route across Europe. Every year, applications are submitted in view of obtaining the certification. Two main bodies are involved in the certification process. The European Institute of Cultural Routes (EICR), the technical agency set up as part of a political agreement between the Council of Europe and the Grand-Duchy of Luxembourg, carries out the first assessment of the application and the follow-up of the independent experts' reports. The Enlarged Partial Agreement on Cultural Routes (EPA), supported by member states and mainly represented by Ministries of Culture, Tourism and Foreign Affairs, takes the final decision on the award certification each year during its April meeting (Council of Europe & European Union 2020).

The certification of a cultural route entails a host of benefits for the countries and regions concerned, for instance, participation in a wide transnational network, access to specialized researchers and experts, use of the label Cultural Route of the Council of Europe and use of the Council of Europe logo. This allows for the exchange of ideas and development of synergies with other European cultural routes, ensures its promotion and broadens its recognizability, thus attracting large numbers of visitors. It also facilitates access to various sources of European, national and regional funding.

7. Business Activity

In Greece, a lot of travel and tourism agencies specializing in religious tourism organize and offer special tour packages and faith-based tours, inviting tourists to follow in the footsteps of Apostle Paul. Most of these agencies are certified by the Church of Greece. In order for a Greek travel agency to be certified by the Church of Greece, it should express and implement the objectives, the actions and the activities of the Synodical Office of the Church of Greece, established in Regulation no. 281/2015. In particular, the proposed pilgrimage tours should showcase the religious, historical, cultural and ecological wealth of the churches and monasteries and religious-cultural heritage dispersed throughout Greece.

The tours usually cover a part of the full itinerary and only last a few days because tourists need to have a lot of available time in order to travel the whole itinerary. Biblical tours organized by tour operators are often focused on tourists visiting the places connected with Paul's preaching, with no activities in the program.

In contrast, in the excursions organized by the Greek Holy Metropolises (e.g. an educational youth pilgrimage excursion organized by the Holy Metropolis of Kitros, Katerini, and Platamon[17]), the program went beyond visits to cultural sites and monuments. It also included various activities such as pilgrimage, guided tours, watching documentaries on St. Paul, having lunch and dinner in accordance with the Greek orthodox tradition and attending Mass. The route is not just a geographical itinerary but a spiritual experience based on existing and modern elements. The pilgrimage excursion program meets the demands of those who wish to have an authentic experience, genuine feelings and spiritual challenges, which are also the pilgrim's goals.

The goal of these organized pilgrimage tours is for its participants to become integrated into the natural environment, comprehend the local culture and get away from the hustle and bustle of everyday life. The route helps visitors release their creative potential through their participation in locally organized educational activities.

With regard to the development of entrepreneurship in this field serious problems arise due to the plethora of public and private bodies and agencies involved in monument management, and due to the fact that there is no central management agency. More specifically, researchers highlight the absence of an integrated tourism product that can be used to promote the route. For example, Vasilios Mygdalis (2015) conducted interviews with representatives from the agencies and bodies involved in the development of the route in northern Greece, and concluded that there is no coordinated and concerted effort at a local, regional and national level for the development of religious tourism owing to deficient communication and cooperation among the stakeholders involved. He also underlines the fact that the apostolic mission of the Holy Metropolises is exclusively apostolic, which does not allow touristic valorization of the involved monuments and sites. Further, entrepreneurship in the field of religious tourism is hindered by the shortage or even nonexistence of the appropriate infrastructure, the inadequate training of the workers, the lack of an institutional framework dedicated to the route and the inefficient or nonexistent promotion of the Cultural Route.

The award of official European classification to the Footsteps of Apostle Paul route in the future and the cooperation this should engender among the stakeholders could resolve many of the aforementioned problems.

8. Conclusion

Cultural routes link regions with similar historical or physical features. They refer to different periods of history and relate to the roads of conquerors, the maritime and land trade roads, monuments and areas associated with specific communities. The cultural routes connecting classical antiquity with Christianity in Greece attract thousands of visitors with various motivations. A typical example is the discussed route Along the Footsteps of Saint Paul, in Greece, which features the historical and cultural-religious relations developed between single sites, historical cities and areas focusing on Paul and his preaching. Regardless of the length and the duration of the trail, which revives Paul's missionary journeys, this route encourages visitors to make a mental journey through place and time, by directly relating to history, culture, faith, natural environments and local people. The route might also raise awareness of the religious-cultural tradition it is based upon as well as its heritage. The touristic experience is enhanced by the complexity and the diversity of the route's visitor experiences.

This resurgence of the religious-cultural route of Apostle Paul allows for a revival of collective memory and an opportunity for the local communities involved to restore and reinforce their historical, religious-cultural, environmental and economic resources. In addition, it may serve as a basis for intercultural dialogue in the Mediterranean Basin, a region full of intense religious, cultural and political conflicts.

List of Abbreviations

CIIC	International Committee on Cultural Routes
CoE	Council of Europe
GNTO	Greek National Tourism Organization
ICOMOS	International Council on Monuments and Sites
MoU	Memorandum of Understanding
NGOs	Non-Governmental Organizations
UN	United Nations
UNESCO	United Nations Educational Scientific and Cultural Organization
UNWTO	United Nations World Tourism Organization

Notes

1. EICR was set up in 1998 as part of a political agreement between the Council of Europe and the Grand-Duchy of Luxembourg (Ministry of Culture, Higher Education and Research). It provides advice and assistance to routes networks and hosts visits by project managers, researchers and students (EICR 2018).
2. 'Cemeteries are part of our tangible heritage, for their sculptures, engravings, and even for their urban planning. They are also part of our intangible heritage. Both make up the funerary heritage. Cemeteries provide unique settings where to find part of our historical memories. They are places where to remember periods of local history that we have the duty to preserve and transmit to future generations' (European Cemeteries Route 2018).
3. The Routes of the Olive Tree were inaugurated in 1998. They were nominated 'International Cultural Itinerary of intercultural dialogue & sustainable development' in 2003 by UNESCO, and 'Great European Cultural Itinerary' in 2006 by the Council of Europe. The Routes are based on the theme of the olive tree, a universal symbol of peace and dialogue. Their aim is to promote the Mediterranean olive-oil producing regions and products, by valorizing the culture enclosing them. The Cultural Foundation 'Routes of the Olive Tree' is in Kalamata, Greece.
4. The Routes of the Orthodox Monasteries refer to the architectural and cultural wealth of the Orthodox monasticism, a cooperation project between Greece, Albania, Bulgaria and FYROM. The aim of the project is to promote the shared cultural heritage through the Orthodox monuments and worship, the revival and the preservation of historical data, mutual understanding and the mapping of the paths which led monks and travelers to the monasteries located along the Via Egnatia. The Via Egnatia was built to facilitate communication between East and West, between the Adriatic, the Aegean and the Hellespont (Black Sea), Rome and Byzantium-Constantinople. One of the deliverables of the project was the publication of three cultural-tourist guides 'Monasteries of the Via Egnatia' (2 vols, in Greek and English) and 'Monasteries of the

Aegean islands' (1 volume, in Greek and English), which was funded by the INTERREG II Program of the European Commission (Ministry of Culture of Greece 1999a, 1999b, 1999c).

5. *Ekklēsia* (Church), in Christian theology means both a particular body of faithful people and the whole body of the faithful. The term also means assembly, congregation, council.
6. Ancient Galatia was an area in the highlands of central Anatolia in modern Turkey. Galatia was named for the immigrant Gauls from Thrace, who settled here and became its ruling caste in the third century BCE, following the Gallic invasion of the Balkans in 279 BCE. It should not be confused with Gaul, a region of Western Europe during the Iron Age that was inhabited by Celtic tribes, encompassing present day France, Luxembourg, Belgium, most of Switzerland, Northern Italy, as well as the parts of the Netherlands and Germany on the west bank of the Rhine.
7. Kali Limenes (meaning 'Fair Havens') is a village and port, located 82 kilometers south-west of the city of Heraklion, Crete. The small island of Agios Pavlos ('Saint Paul') lies at the port's entrance. Agios Pavlos bay is where Saint Paul allegedly sought shelter during a violent storm (Tzerpos n.d.: 7). The bay owes its name to the white chapel standing on the beach which is dedicated to Saint Paul.
8. The monument is entitled 'The arrival of the Apostle Paul in Neapolis of Philippi', a work of the Greek painter Vlassios Tsotsonis and the Italian mosaic-maker Pino Pastorutti (Dimofelia 2017).
9. The Archaeological Site of Philippi is also a listed historic monument in Greece (Ministerial Decision No. 15794/19-12-1961, Gov. Gaz. 35 B/2-2-1962).
10. Ministerial Decision No. 23085/738/25-8-1948 - Gov. Gaz. 10/23-9-1948.
11. Ministerial Decision ΥΠΠΟ/ΓΔΑ/ΑΡΧ/Α1/Φ43/51641/3084/5-10-2001 – Gov. Gaz. 1387/B/22-10-2001.
12. Some of the celebrations are held at the Vima tribunal facing the agora in ancient Corinth, where the city's bishop holds vespers (Holy Metropolis of Corinth, Sicyon, Zemenon, Tarsos and Polyphengos 2017).
13. The run was organized on 21 May 2017 by the Kavala branch of the Hellenic Athletics Federation, under the aegis of the Holy Metropolis of Philippi, Neapolis and Thassos and the Hellenic Athletics Federation, in cooperation with the Municipality of Kavala, Dimofelia and the Regional Unit of Kavala (SportEvent 2017).
14. For the Santiago de Compostela Pilgrim Routes, separate multilingual maps for the nine main routes have been published, which are distributed to pilgrims and tourists. These maps contain information on the monuments, accommodation and other services along each route as well as the profile of each route (e.g. altitude, degree of difficulty). They also contain recommendations on how pilgrims should behave during the pilgrimage.
15. A Metropolis or Metropolitan Archdiocese is a see or city whose bishop is the Metropolitan of a province. Metropolises, historically, have been important cities in their provinces. In the Eastern Orthodox Churches, a metropolis is

a type of diocese, along with eparchies, exarchates and archdioceses. In the Churches of Greek Orthodoxy (Church of Greece, Greek Orthodox Archdiocese of America, etc.), every diocese is a metropolis, headed by a Metropolitan.

16. The European Union (EU) values are common to the EU countries in a society in which inclusion, tolerance, justice, solidarity and non-discrimination prevail. These values (human dignity, Freedom, democracy, equality, rule of law and human rights) are an integral part of the European way of life.

17. The excursion took place on 18 February 2018 under the escort of the Very Reverend the Metropolitan of Citrus, Katerini and Platamon, Georgios (Poimin 2018).

References

Antunes, A., S. Amaro and C. Henriques (2017), 'Motivations for Pilgrimage: Why Pilgrims Travel El Camiño de Santiago', *International Religious Tourism and Pilgrimage Conference*, Armeno-Orta Lake, Italy, 28 June–1 July, http://arrow.dit.ie/irtp/2017/visitor/1/ (accessed 5 July 2017).

Athina 9,84 (2017), 'The Steps of Saint Paul "Open" Tourism from Nigeria', e-newspaper 20 July (in Greek), http://www.athina984.gr/2017/07/20/ta-vimata-tou-apostolou-pavlou-anigoun-ton-tourismo-apo-ti-nigiria/ (accessed 25 July 2017).

Battaini-Dragoni, G. (2015), *Speech at the Council of Europe Cultural Routes Annual Advisory Forum*, Aranjuez, Spain, 29 October, http://www.coe.int/en/web/deputy-secretary-general/speeches/-/asset_publisher/Gt0K7o1XnY6l/content/council-of-europe-cultural-routes-annual-advisory-forum (accessed 15 March 2017).

Belezos, C. J. (ed.) (2009), 'The Apostle Paul and Corinth: 1950 Years after the Writing of the Epistles to the Corinthians', *International Scholarly Conference Proceedings*, Vols. 1 and 2, Athens: Psichogios Publications.

Church of Greece (2015), *Regulation no. 281/2015* 'On the establishment and operation of the Synodical Office of Pilgrimage Tours of the Church of Greece' (Gov. Gazette 2/A/13-1-2016) (in Greek).

Constitution of Greece, The (1978), *As Revised by the Parliamentary Resolution of May 27th, 2008 of the VIIIth Revisionary Parliament*, Athens: Hellenic Parliament, https://www.hellenicparliament.gr/UserFiles/f3c70a23-7696-49db-9148-f24dce6a27c8/001-156%20aggliko.pdf.

Council of Europe (2010), *Resolution CM/Res(2010)53, Establishing an Enlarged Partial Agreement (EPA) on Cultural Routes*, adopted by the Committee of Ministers on 8 December 2010, at its 1101st meeting of the Ministers' Deputies, https://search.coe.int/cm/Pages/result_details.aspx?ObjectID=09000016805cdb50 (accessed 5 July 2017).

Council of Europe (1994), *Via Francigena*, https://www.coe.int/en/web/cultural-routes/the-via-francigena (accessed 4 February 2017).

Council of Europe (2005), *Saint Martin of Tours Route*, https://www.coe.int/en/web/cultural-routes/the-saint-martin-of-tours-route (accessed 4 February 2017).

Council of Europe (2013), *Resolution CM/Res (2013)67 Revising the Rules for the Award of the 'Cultural Route of the Council of Europe' Certification* (Adopted by the Committee of Ministers on 18 December 2013 at the 1187bis meeting of the Ministers' Deputies), https://search.coe.int/cm/Pages/result_details.aspx?ObjectId=09000016805c69fe (accessed 6 August 2018).

Council of Europe (2017), *Enlarged Partial Agreement on Cultural Routes*, https://www.coe.int/en/web/culture-and-heritage/cultural-routes (accessed 5 January 2019).

Council of Europe (2018a), *'Cultural Route of the Council of Europe' Certification Conditions for Obtaining the 'Cultural Route of the Council of Europe' Certification*, https://www.coe.int/en/web/cultural-routes/applications-certification (accessed, 21 August 2020).

Council of Europe (2018b), *Cultural Routes*, https://www.coe.int/en/web/cultural-routes/by-theme (accessed 3 February 2018).

Council of Europe (2020), *St Paul's footsteps candidate Cultural Route of CoE*, 24/04/2020, https://www.coe.int/en/web/cultural-routes/applications-certification (accessed 21 August 2020).

Council of Europe & European Union (2020), *How to be Certified 'Cultural Route of the Council of Europe'?*, https://pjp-eu.coe.int/en/web/cultural-routes-and-regional-development/certification-guidelines#{%2245462558%22:[1] (accessed 03 February 2020).

Del Chiappa, G. and F. Fortezza (2015), 'Weeding Base Tourist Development: Insights from an Italian Context', *Advances in Culture, Tourism and Hospitality Research*, 10: 61–74.

Despotis, S. S. (2009), *The Apostle Paul in Athens*, Athens: Athos Editions.

Dimofelia (2017), *The Monument of the Apostle Paul*, Municipality of Kavala, Public Benefit Organization, (in Greek), http://www.kavalagreece.gr/tourismos/touristikes-plirofories/aksiotheata/mnimio-apostolou-paulou/ (accessed 4 February 2018).

EICR (2018), *European Institute of Cultural Routes*, https://www.coe.int/en/web/cultural-routes/european-institute-of-cultural-routes (accessed 3 February 2018).

European Cemeteries Route (2018), *About Cemeteries Route*, https://cemeteriesroute.eu/about-cemeteries-route.aspx (accessed 3 February 2018).

Fisher, R. (1992), *Who does What in Europe?*, London: Arts Council.

Geraki (2006), Map Showing the Routes of Apostle Paul's Journeys, *Wikipedia*, https://el.m.wikipedia.org/wiki/%CE%91%CF%81%CF%87%CE%B5%CE%AF%CE%BF:Apostle_Paul%27s_journeys.svg.

Geanakoplos, D. J. (1965), 'Church and State in the Byzantine Empire: A Reconsideration of the Problem of Caesaropapism', *Church History*, 34 (4): 381–403.

Greek National Tourism Organization (2003), In the Footsteps of Paul the Apostle in Greece, Athens.

Goutzioudis (n.d.), *The New Testament, Acts of the Apostles*, Apostoliki Diakonia of the Church of Greece, 10 January 2001, Athens: To Vima (in Greek).

Hadjinikolaou, E., G. Zirinis and M. Sofikitou (2015), *For a Cultural Route, Diazoma*, http://www.diazoma.gr/site-assets/1.-%CE%9A%CE%B5%CE%AF%CE%BC%CE%B5%CE%BD%CE%BF-%CE%B2%CE%AC%CF%83%CE%B7%CF%82-%CE%B3%CE%B9%CE%B1-%CE%94%CE%B9%CE%B1%CE%B4%CF%81%CE%BF%CE%BC%CE%AD%CF%82.pdf (accessed 17 August 2018).

Holy Metropolis of Veria, Naoussa and Campania (2017), *International Scientific Conference 'St Paul and the Philosophers'* (in Greek), https://www.imverias.gr/index.php/payleia/62-kv-payleia-2016/467-oloklirothike-to-22o-diethnes-epistimoniko-synedrio-apostolos-pavlos-kai-filosofoi-foto (accessed 15 March 2020).

Holy Metropolis of Corinth, Sicyon, Zemenon, Tarsos and Polyphengos (2017), *11th Corinthou Pavleia 2017* (in Greek), http://www.imkorinthou.org/index.php/pavleia/105-ia-korinthou-pavleia-2017/3237-ia-korinthou-pavleia-2017-programma-ekdiloseon (accessed 5 July 2017).

Holy Metropolis of Thessaloniki (n.d.), *Towns-stopovers during the missionary journey of Paul the Apostle in Greece, Cyprus, Malta and Southern Italy*, translated by I. Sideridou, Kentro Paideias kai Politismou, Vocational Training Center, Thessaloniki, p. 7 (in Greek).

Kathimerini, I. (2017), 'Following the Steps of Saint Paul the Apostle', 15 April, http://www.kathimerini.gr/762906/article/epikairothta/ellada/akoloy8wntas-ta-vhmata-toy-apostoloy-payloy (accessed 16 March 2017) (in Greek).

ICOMOS (2008 [2003]), *The ICOMOS Charter on Cultural Routes*, Elaborated and revised by the International Scientific Committee on Cultural Routes (CIIC) of ICOMOS. Version approved by the Executive Committee of ICOMOS (Pretoria, October 2007) and definitively ratified by unanimity at the General Assembly (Quebec, October 2008), http://www.icomos-ciic.org/ciic/Charter_Cultural_Routes.pdf (accessed 14 February 2018).

Ioannides, Th. (2009), *The Anthropos and the World according to St. Paul*, Athens: Pournaras Editions (in Greek).

Karavasili, M. and E. Mikelakis (1999), 'Industrial Tourism: Promoting the Industrial Heritage in the Region of Sterea Ellada through a Grid of Special Interest Cultural Routes', *Technologia*, 9: 50–2 (in Greek).

Major, B., F. McLeay and D. Waine (2010), 'Perfect Weddings Abroad', *Journal of Vacation Marketing*, 16 (3): 249–62.

Majdoub, W. (2010), 'Analysing Cultural Routes from a Multidimensional Perspective', *AlmaTourism*, 2: 29–37.

Meletios His Eminence, Metropolitan of the Holy Metropolis of Nikopolis and Preveza (2015), *Saint Paul and Nikopolis*, 26 June (in Greek), https://www.pemptousia.gr/2015/06/o-apostolos-pavlos-stin-nikopoli/ (accessed 5 July 2017).

Metallinos, G. (1999), *Saint Paul*, Holy Metropolis of Cephalonia (in Greek), http://www.new.imk.gr/el/showcat.asp?CatID=5&scatid=25 (accessed 18 March 2017).

Ministerial Decision No. 15794/19-12-1961 (Gov. Gazette 35/B/2-2-1962) (in Greek).

Ministerial Decision no. 23085/738/25-8-1948 (Gov. Gazette no. 10/23-9-1948) (in Greek).

Ministerial Decision no. ΥΠΠΟ/ΓΔΑ/ΑΡΧ/Α1/Φ43/51641/3084/5-10-2001 (Gov. Gazette no. 1387/B/22-10-2001) (in Greek).

Ministry of Culture of Greece (1999a), *The Routes of Orthodox Monasteries. Monasteries of the Via Egnatia*, Vol. 1, Athens.

Ministry of Culture of Greece (1999b), *The Routes of Orthodox Monasteries. Monasteries of the Via Egnatia*, Vol. 2, Athens.

Ministry of Culture of Greece (1999c), *The Routes of Orthodox Monasteries. Monasteries of Aegean Islands*, Vol. 3, Athens.

Moira, P. (2020), *Religious Tourism and Pilgrimage: Politics – Management – Sustainability*, Athens: Fedimos (in Greek)

Mygdalis, V. (2015), 'Opportunities for the Development of Religious Tourism in Eastern and Central Macedonia: "The Footsteps of Saint Paul"', Master's diss., Chios: University of the Aegean (in Greek).

Mylonopoulos, D. (2001), *The Constitution of Greece 1975/1986/2001: A Legal Vetting Approach of the Revision of the Year 2001*, Athens: Stamoulis Publications (in Greek).

Oikonomou, Ch. (2008), *The Summons and the Beginnings of the Missionary Action of St. Paul*, Athens: Pournaras Editions (in Greek).

Poimin (2018), *Youth Pilgrimage Excursion of the Holy Metropolis of Kitros 'In the Footsteps of Paul the Apostle'*, http://poimin.gr/proskinimatiki-ekdromi-neon-tis-i-m-kitrous-sta-vimata-tou-apostolou-pavlou/ (accessed 8 July 2018).

Proto Thema (2012), 'Couples from All Over the World Go to the Island of Rhodes to Get Married!', electronic newspaper, 12 October (in Greek), https://www.protothema.gr/greece/article/229505/sth-rodo-gia-na-pantreytoyn-erxontai-zeygaria-apo-olo-ton-kosmo-/ (accessed 15 February 2020).

SportEvent (2017), 'Run "In the Steps of Saint Paul"', http://sportevent.gr/index.php/calendar/old-calendar/item/10457-agonas-dromou-sta-vimata-tou-ap-paylou-2017 (accessed 4 February 2018).

To Vima (2001), *The New Testament: Acts of the Apostles*, Apostoliki Diakonia of the Church of Greece, 10 January, Athens (in Greek).

Tzerpos, V. (n.d.), 'Other Cities Stopovers of St. Paul in Greece', in Greek National Tourism Organization, *In the Footsteps of St. Paul the Apostle in Greece* (in Greek).

UNESCO (2016), *Archaeological Site of Philippi*, https://whc.unesco.org/en/list/1517 available on line 17/8/2018.

UNWTO (2015), *Global Report on Cultural Routes and Itineraries*, Affiliate Members Report: Volume twelve, UNWTO, Madrid, Spain, http://cf.cdn.unwto.org/sites/all/files/pdf/global_report_cultural_routes_itineraries_v13.compressed_0.pdf (accessed 17 August 2018).

UNWTO (2019), *Compendium of Tourism Statistics Dataset [Electronic]*, UNWTO, Madrid, updated 20 November 2019, https://www.e-unwto.org/doi/abs/10.5555/unwtotfb0300010020142018201911 (accessed 14 March 2020)

Part III

Heritagization

Chapter 8

BIBLE MUSEUMS

Crispin Paine

Introduction

It is hardly surprising that there are a lot of Bible museums.[1] The Bible is surrounded by superlatives: most published, most translated, most owned and so on; above all it has, as Christopher De Hamel puts it, an 'emotionally charged place in popular imagination' in large swathes of the world (2001: vii).

This chapter looks at some of the many different ways Bible museums present the Bible to tourists and other visitors. I shall argue that 'Bible Museum' is a useful term for a wide variety of attractions from the huge to the tiny, from archaeological exhibits explaining the Bible's background to bibliophiles' exhibits of Bible collections, and from exhibitions focused on one great illuminated manuscript to those showing many Bible-based works of art. What all Bible museums share is their basis in the Bible, as a living text, as a historical phenomenon or as a physical thing. I therefore define 'Bible Museum' in the simplest possible way: as a museum that presents some aspect of the Bible.

Recent scholarship has examined the great variety of ways in which Bibles can be used.[2] They divide firstly into (a) how the text is read, understood and interpreted, and (b) how the physical object is used. The Bible's text has been used in many ways throughout its known history: as history, as allegory, as instructions for living and as political contract and – for many – as the supreme source of spirituality and revelation of God's nature and purposes. The physical object, whether manuscript or printed book, is also used or studied in a great number of ways: as venerated icon channeling the divine, as work of art or bibliographic curiosity, as historic specimen or liturgical tool, as personal memento or good luck charm. Understanding the Bible as text and understanding it as an object interplay continually, and recognition of the two approaches has been powerfully encouraged by the appearance of digital Bibles, available on computer, e-book reader or smartphone (Rakow 2017).

Definitions of 'museum' all stress three things: building up and caring for a collection, using that collection for the education and enjoyment of the public, and making sure that the collection is preserved for future generations to enjoy and to use (Ambrose and Paine 2018: 11). Though they may emphasize different aspects, all Bible museums have one underlying purpose. They aim to use their collection to help people understand the Bible: what it says, how it happened and what its impact has been. I want in this chapter, though, to examine some of the varied approaches Bible museums take to this underlying purpose. These approaches grow out of the museums' (usually their founders') different approaches to the Bible itself; they result in different kinds of collections, different ways of deploying those collections and different levels of concern for their long-term preservation.

Museums and the Collector's Bible

A striking number of Bible museums have grown out of individuals' collections of Bibles; there is a big community of Bible collectors, sometimes very scholarly.[3] For many collectors their principal motive is a love of old books, for others a devotion to the Bible; others again love both. One of the greatest collections of Bibles is that of Martin Schøyen, based in Oslo, itself just a part of what is billed as the largest private manuscript collection in the world. Other great collections are clearly prompted as much or more by devotion to the Bible as the Word of God as by bibliophile enthusiasm; such private collections not infrequently pass into academic library collections or become museums open to the public. The Scriptorium is one of the principal attractions of the Holy Land Experience, the well-known Evangelical Christian theme park in Orlando, Florida. It is based on what is said to be the second largest collection of Bibles – printed books and manuscripts – in the world, built up by the late investment banker Robert van Kampen, who was both one of the richest men in the United States and a passionate Evangelical who wrote widely on fundamentalist Christian eschatology (Carroll 2007). The foundation,[4] established by Van Kampen's widow, lent his 25,000-object collection (said to be insured for $40m [Carroll 2007: 236]) to the Holy Land Experience in 2002. It is displayed there in a very lively manner, with much use of laser lights, animatronic figures and sound effects. The tour takes almost an hour, through fourteen themed rooms, each of which reconstructs a scene in the story of the Bible, including the Library of Alexandria, a bindery in Constantinople, John Wycliffe's study, William Tyndale's workshop, John Bunyan's prison cell, the Metropolitan Tabernacle in Victorian London and a prairie church in Puritan America. It is probably seen by most of the roughly 250,000 visitors a year to the theme park.

A more modest example of a Bible museum based on a personal collection opened in 2009 at St. Arnaud, three hours' drive northwest of Melbourne, Australia. Built up by Bible student Ellen Reid and her daughter Jean (who also collect maps and breed butterflies), the museum displays over two thousand Bibles of every description, plus models and artifacts. As Ellen Reid puts it, 'The museum aims to raise awareness of the most published book in the world, and to educate people about the book itself. Australia is a very secular country that does not have the religious heritage of the European countries.' They funded it by selling their Melbourne home and buying a semi-derelict property in a remote country town. The museum attracts some 2,500 visitors a year,[5] a good number for such a specialist museum in such a location. More typical of Bible-collection based museums is perhaps the collection of Kansas university professor Peter McGraw, whose collection of Bibles and Bible-related things formed the basis of the Bible Museum in the Great Passion Play Theme Park in Eureka Springs, Arkansas. Leading out of the shop, it is seen by many if not most of the park's annual 50,000 visitors.

Another Bible museum based on a personal collection is the Museum of the Book in London's Docklands.[6] This museum is a series of display cabinets around the walls of a 1960s Baptist chapel, which now mainly serves the local Bangladeshi community. The museum was set up by the Rev. David Smith, formerly the pastor of a chapel in fashionable Kensington, and for many years a collector of and dealer in Bibles and related items. He was largely responsible for assembling the Van Kampen collection, and sold half his own collection to Steve Green, the multi-millionaire founder of Washington's Bible Museum. He is still collecting – or rather amassing – energetically, today mainly on behalf of the museum's owner, the Salmon Lane Mission, which now has a very large collection. At the time of my visit the museum was in the process of replacing its displays on the history of the Bible, celebrating the Reformation centenary, with an exhibition on morals and ethics. The museum concentrates on group visits, which include a personal two-hour tour; Dr. Smith has a special skill in choosing evocative objects from his collections to make his points and a very remarkable ability to present his message – a rare combination of the skills of the curator and the preacher.

Figure 8.1. Behind the scenes at a Bible-collection museum.
Photo by author.

The collector/dealer's museum is at its most prominent in The Bible Museum at Goodyear, Arizona. This museum is part of Greatsite.com, the 'World's Largest Dealer of Rare & Antique Bibles', who will sell you an original leaf from the 1540 Great Bible for $295, or one from the 1455 Gutenberg Bible for $100,000.

Another source of the collection-of-Bibles type of Bible museum is missionary organizations. In 1804, a group of people including William Wilberforce, founded the British and Foreign Bible Society to print and distribute affordable Bibles. Today there are over 140 Bible Societies throughout the world, many of which have scholarly libraries and some of which have museums. Thus, The Bible Society of India was founded in Calcutta in 1811 with an emphasis on translating the Bible into Indian languages. In 2004 it set up The Bible World in its Bangalore office block, a museum which receives some ten visitors a day, mainly Christians. It displays hundreds of Bibles, including many translations published by the Society, in rows of showcases, to 'spread the Word and show how the Bible came to us'. There are also Bible-related postage stamps, a model of the Holy Land, such curiosities as a complete Bible on one page, and computer interactives (see Fig. 8.2).[7]

Figure 8.2. The Bible World, Bangalore, India. Photo by author.

Museums and the Archaeologist's Bible

Biblical archaeology is a special 'armchair' variety of general archaeology. The biblical archaeologist may or may not be an excavator himself, but he studies the discoveries of the excavations in order to glean from them every fact that throws a direct, indirect or even diffused light upon the Bible. (…) Yet his chief concern is not with methods or pots or weapons in themselves alone. His central and absorbing interest is the understanding and exposition of the Scriptures. (Wright 1957: 17)

Thus begins C. Ernest Wright's classic work of 1957, *Biblical Archaeology*. Since then a good many museums have been set up to tempt the biblical archaeologist out of his or her armchair, almost all of them aiming to promote 'the understanding and exposition of the Scriptures'. They do so by presenting the archaeology of those lands and times in which the Bible was created – effectively from the eighth century BCE to the first century CE. Most use models, replicas, dioramas and the many other display techniques of a modern museum to tell the story, and such museums can trace their ancestry back for centuries, as Annabel Wharton describes in her wonderful 'Selling Jerusalem' (2006). A modern example, opened in 2010, is the Bibleworld Museum and Discovery Centre in Rotorua,

New Zealand. This uses models, replicas and original artefacts in its Old Testament and New Testament galleries. Its 'Hands-on Area' gives visitors the chance to explore an interactive relief map of the lands Jesus knew, to blow the ram's horn shofar and play the ten-stringed lyre, to dress up as a Roman soldier or to write their names in hieroglyphs or cuneiform.[8]

A few Bible museums use immersive techniques to give visitors a full-blooded experience of (and thus an empathy for) life in biblical times, like those more often found in theme parks (Paine 2019). The most notable examples are perhaps the Museum of the Bible in Washington, D.C., where on the third floor visitors can walk around Nazareth in the time of Christ (see Fig. 8.3), and the Creation Museum in Kentucky, where visitors can explore the Garden of Eden or listen to Noah and his workers talking about the Ark (Summers 2017: 156; Ham 2008). They use a mixture of animatronic figures and human actors in carefully reconstructed environments (see Fig. 8.4).

Figure 8.3. Building the set for 'The World of Jesus of Nazareth', Museum of the Bible, Washington, D.C. Photo by author.

Figure 8.4. Adam in the Garden of Eden, Creation Museum, Kentucky.
Photo by author.

A museum that takes a quite different approach to either of those, while still making strong use of archaeology, is the *Bibelhaus Erlebnis* Museum (Experiential Bible House Museum), just south of the river Main in the center of Frankfurt, Germany (Schefzyk 2016). Opened in 2003 in a modest former church building, the museum saw 25,000 visitors in 2017 and puts a strong emphasis on school parties. It is run by the *Frankfurter Bibelgesellschaft*. On the top floor, devoted to the Old Testament, a rather modest space contains little but a Bedouin-style tent, presumably for school parties to sit in, plus a model of the Temple and an interactive or two. Much more engaging is the more recent ground-floor display on the New Testament, entitled 'Judea 2000 Years Ago' and imaginatively conceived and well designed. The display is striking for not focusing on Jesus, but on the world in which he lived and preached. It is organized along some fifteen themes, such as Sacrificing (the Temple), Travelling (pilgrimage to Jerusalem), Relishing (the upper classes in first-century Judea), Conquering (Roman occupation), Worship (the Imperial cult and its analogies with early Christianity), Revolting (Judean revolt and destruction of the Temple), Celebrating (boy's passage to manhood and girl's passage to womanhood) and Life at the Lake (Lake Galilee). The second half of the display is dominated by a full-size replica of a Galilean fishing boat. Each exhibit includes well-written text (with English translation), a few objects (including 270 original and reproduction objects lent under a ten-year agreement by the Israeli Antiquities Administration), reproductions, video-screens, recordings and interactives.

Another German-speaking Bible museum that concentrates even more firmly on children is *Bibelwelt* (Bibleworld) in Salzburg. This is more of a themed adventure playground than a typical museum, as is obvious when you enter the converted church through a gigantic ear. Largely organized around the stories of Jesus and St. Paul, *Bibelwelt* includes an oriental market, a 40m² map of the Mediterranean outlining the journeys of Paul the Apostle, dark passageways and unsteady floors intended to reflect Jesus' 'tale of woe', and a one-arm bandit: the 'Mercy Slot-Machine'. The museum attracts over 7,000 visitors a year, but notably last year drew 239 school groups.[9]

Thus the 'archaeologist's Bible museum' can stretch from a fairly academic object-focused exhibit to something more like a theme park.

Nor need such museums be Christian in focus, though most are. In Brooklyn is The Living Torah Museum, which 'is dedicated to helping those who study the Torah (Bible) get a better understanding of the items mentioned in the Torah by exhibiting ancient artifacts'. It extends that both by including displays on Torah scholars ('Great Torah Personalities'), and through a sister museum, Torah Animal World, a taxidermy museum in the Catskills.

In the Holy Land itself, however, a Bible museum becomes a local history museum. The Bible Lands Museum in Jerusalem began as the personal collection of Elie Borowski, a Polish Jewish scholar who transferred it from Canada to Israel, and in 1992 opened it as a substantial, professional and lively museum dedicated to telling the story of the Bible's peoples.[10] Besides the permanent galleries, the museum holds regular Bible-related exhibitions; in 2018 for example 'Jerusalem in Babylon: New Light on the Judean Exiles', built on the al-Yahuda tablets, told the story of the destruction of the first Temple and the exile in Babylon.

Museums and the Politician's Bible

Bible museums can play an important role in those lands where religion, nationalism and cultural politics merge. The Shrine of the Book in Jerusalem was created in 1965 to house the Dead Sea Scrolls. The first scrolls were found by local Bedouin in 1947 and sold to a Bethlehem antique dealer, who in turn sold them to the Archbishop of the Syrian Orthodox Church, who in 1954 sold them on to the Israeli army Chief of Staff turned archaeologist, Yigael Yadin, for the State of Israel. Many more scroll fragments were found in the 1950s and 1960s, and the 1967 Israeli occupation of East Jerusalem and the West Bank ensured that the great majority ended up in the new Israel Museum.

The Shrine's distinctive dome copies the lids of the jars in which the scrolls were originally kept. The contrast between its whiteness and a great black granite wall is held by some to be a reference to the decisive battle between the Children of Light and the Children of Darkness, described in the scroll known as the War Scroll, the final confrontation between good and evil which was to precede the coming of the Messiah.

The visitor enters the Shrine through an underground corridor with displays on life in the Qumran religious community (believed to be the scrolls' origin), before emerging into the dramatic and cave-like domed chamber where the scrolls themselves – or some of them – are displayed. The centerpiece is a reproduction of the seven-meter Scroll of Isaiah, supported by original pieces of that and other scrolls. Displayed in a lower gallery (and sometimes missed by visitors) is the Aleppo Codex, a tenth-century manuscript of the Hebrew Bible with a dramatic history, scribed in Tiberias on the western shore of the Sea of Galilea (Roitman 2006).

The Shrine of the Book supplied the new nation of Israel with a monument that at once declared the antiquity of its people, their association with their newly conquered land and – through a very modern building of international quality – their modernity. That is, it encapsulated the Zionist message, and through the popular association of the scrolls with romance, mystery and Christianity as well as with Judaism, exported that message throughout the Western world and especially to the United States. Indeed, Yigael Yadin (who played a big role in creating the Shrine as well as in acquiring scrolls) said in 1966: 'I think that the Shrine of the Book does an excellent job of public relations for the State of Israel and for Zionism in general'.[11] Still for the Israeli public, as Evan Taylor (2015: 54; see also Knell et al. 2011; Aronsson and Elgenius 2015) puts it,

> one needn't read any site text to read the site as a recentering of the Dead Sea Scrolls in the contemporary nation-state. Situated in full view of the Knesset (the Israeli parliament), (...) the Shrine positions the Scrolls as central documents of the nation.

Bible museums also play a role in the Christian/Jewish/Israeli politics that is so prominent in Evangelical America (and so little understood in the rest of the world). 'Traditional' anti-Semitism has largely given place there to anti-Islamism and a more-or-less-tight adherence to the idea that Israel is destined to play a crucial role in the Last Days, and therefore deserves the support of Christians; Tristan Sturm (2017) has usefully called this 'Judeo-Evangelical nationalism', and shown that for many American Evangelicals their commitment to the State of Israel is greater than their commitment to the United States. While this Christian Zionism can be detected in a number of Bible Museums in the United States, it appears at its most dramatic in the Friends of Zion Museum in Jerusalem itself. First chaired by the late Israeli President Shimon Peres, the museum was opened in 2015 to tell the story of Christian support of Israel, which it does largely through sophisticated audio-visuals. Is it a 'Bible museum'? I suggest it is, because its story is based on one understanding of the story of the Hebrew Bible: the story of God's covenant with Abraham and His fraught relationship with Israel and Judah, and the creation of the modern State of Israel seen as a return and a fulfillment of the Bible's promise.

In other parts of the world and other faith traditions, Bible museums keep neutral over Middle Eastern politics. The Frankfurt Bible Museum, for example, is careful (in the German text at least) to call the country Israel/Palestine. Moreover, if many Bible museums seem to evoke right-wing politics, that certainly is not true of all. The Bijbels Museum in Amsterdam is one of the oldest of Bible museums. Founded by Leendert Schouten (1928–1905), a clergyman of the Dutch Reformed Church, its collections include a model of the Tabernacle, a model of the Temple Mount, an Egyptian collection and archaeological finds from the Holy Land, while its Bible collection includes famous old Bibles, illustrations, paintings and other collected items which together tell the history of the Bible, and its influence on our daily lives, faith and culture. So far, so conventional, but the museum today uses its resources to address real problems within contemporary Dutch society, and does so from a distinctly liberal perspective. The problems of misunderstanding and conflict between the overwhelmingly secular Dutch society, Muslim immigrants and Jewish communities are addressed through exhibitions, events and an active education programme (van der Meer 2010: 132).[12]

Museums and the Historian's Bible

Throughout the Christian world, and in very many minority communities, whether recently converted or anciently Christian or Jewish, the Bible has played a crucial role in literature, daily life, education and family life. Oddly though, few Bible museums foreground the cultural history of the Bible. One that certainly does is Washington's Museum of the Bible. There the whole second floor is dedicated to the Bible's 'impact', with displays on the Bible in America, the Bible in the World, the Bible Now and a Virtual Reality fly-through of buildings in Washington with Bible links, while the fourth floor has 'The History of the Bible Artifacts' and 'Drive Thru

History of the Bible Theater'. Generally, though, Bible museums seem to leave the historical task to 'secular' history museums. One example from the thousands worldwide might be the Don Yoder Collection of German-language Bibles at the Pennsylvania German Cultural Heritage Center; this is a museum at Kutztown University, Pennsylvania, devoted to the story of the local 'Pennsylvania Dutch' community (Yoder 2016).

Few Bible museums, too, seem to address the use made of the Bible by oppressed groups, like slaves in the United States or first nation peoples in settler countries. Though in 2018 the Bible Museum held an exhibition on 'The Slave Bible', such issues tend to get left to secular museums such as the African American or the American Indian museums in Washington, D.C.

Another more modest Bible museum that gives equal weight to history as to archaeology is the Bible History Exhibits in Pennsylvania County, Philadelphia – aimed no doubt at tourists attracted to the area by the lifestyle of the Amish people. This is a very personal museum housed in a small bungalow on the main road. The simple displays are mostly of museum reproductions of artifacts, inscriptions and manuscripts, collected over the past twenty years and carefully chosen to tell the story first of the Hebrew Bible, then of the New Testament and then of the Bible's impact. In the garden is a modest replica of a Palestinian tomb and an olive press. The one-hour tour is led by Dr. Stephen Myers, who describes a selection of the exhibits, following broadly the story of the Bible (see Fig. 8.5).

Figure 8.5. Bible History Exhibits in Pennsylvania County, Philadelphia.
Photo by author.

Some museums use temporary exhibitions to look at the impact of the Bible, Hebrew or Christian. When I visited in February 2018, the Frankfurt Bible Museum was holding an exhibition on the worldwide spread of the Bible to mark the 500th centenary of the Protestant Reformation. *Fremde, Heimat, Bibel* (Abroad, Home, Bible) was, compared with the museum's New Testament gallery described above, comparatively conventional, with a lot of Bibles in different languages, but organized by interviews with cultural figures from different countries. A few months earlier, at Ark Encounter in Kentucky, an exhibition on the spread of the Bible[13] was essentially an exhibition on Christian mission. This too was structured around individuals, focusing on missionaries from fourth-century Frumentius' mission to Ethiopia to nineteenth-century Hudson Taylor's mission to China.

Museums and the Liturgist's Bible

In both Judaism and Christianity, the Bible – the physical object – is treated with the greatest reverence and considerable ceremony. In the synagogue, the Torah is kept in the Ark, the focus of the building; in the ancient Christian churches the Gospels are often kept on the altar and processed with incense and lights to be read to the congregation, as they have been since the first Christian centuries. In many Protestant churches, too, the Bible is afforded a place of honor in the chapel and is closely associated with preaching. Few Bible museums draw attention to this liturgical use of the Bible. Seldom if ever do they display videos of Gospel processions or readings of the Torah, even in the present day, let alone in reenactments of past practices. So, the way that throughout history most people *heard* the Bible, chanted or spoken, is usually ignored (Lampe 1969).

The nearest the museum visitor can get to an appreciation of the honor that has been paid to the Holy Book is through the richness and quality of its production. The medieval cathedral of Winchester in southern England has for 800 years held the great Winchester Bible, created in the city between 1160 and 1175, the largest and finest of English Romanesque illuminated Bibles (De Hamel 2001: 64). It was probably commissioned by Henry of Blois, grandson of William the Conqueror and Bishop of Winchester for over forty years. He almost certainly intended it to be the cathedral's Great Bible, kept on the High Altar and processed as part of the regular liturgy. While the text is the work of a single scribe on the skins of some 250 calves, six different illuminators worked on the brilliantly colored and gilded pictorial initials that start each chapter. The recent conservation of this Bible has given the cathedral the opportunity to display it in a new exhibition center within the cathedral building – which comes near to being one kind of 'Bible museum'.

In contrast to the way many great churches in the West foreground the artistic aspects of their Bible treasures, monks in northern Ethiopia staunchly resist any interference with their ancient stewardship or any attempt to use their treasures to attract visitors. The Abba Garima Gospels were made around 600 CE by the founder of the monastery in which they have probably remained ever since. Some years ago, the French government helped build a small museum outside the monastery wall, so that the Gospels and other treasures could be both well preserved and seen by tourists. The monks were highly suspicious of this 'Bible museum' and soon took the Gospels back into their traditional treasury (Anon. 2018: 83).

We must remember, too, that many of the key Bibles are in the world's great libraries, and many of these are permanently displayed, or regularly appear in temporary exhibitions. The earliest complete Latin Bible, the Codex Amiatinus, normally lives in the Biblioteca Medicea Laurenziana in Florence, but appeared in an exhibition at the British Library in London in the winter of 2018. It was made in a monastery in Northumbria, northern England, in the early eighth century as a gift for the abbot to take to the Pope (De Hamel 2016: 54). The very close relationship between museums and libraries that care for historic books is well known; such libraries can surely sometimes be 'Bible museums'.

Museums and the Curator's Bible

It is not just as ritual objects that Bibles have been treated as very special. The Codex has always been, as Peter Horsfield and Kwabena Asamoah-Gyadu put it (2011) 'an important identity marker for Christianity as a religion of the book rather than just the text', and the Book has been used in a great variety of ways. Bibles appear in museums – often in secular museums – for a wide variety of sometimes quite odd reasons.[14] Merseyside Maritime Museum in Liverpool has a Bible on which the captains of ships arriving in the River Mersey were obliged to swear that their ships were free of disease; the Bible was conveyed from the Revenue Cutter to the visiting ship by rope. Many museums have examples of the famous 'Bible that stopped a bullet', from the **Sam Houston Memorial Museum in Texas** to the Australian War Museum in Canberra.

One common reason Bibles get preserved is their personal association. Thus, the William Booth Birthplace Museum in Nottingham, England, displays the Bible used by the founder of the Salvation Army in his early years preaching on the city's streets. This is clearly the right place to preserve it, but sometimes association Bibles seem to enter collections as mere curiosities. The Museum of the Book in London shows Bibles that once belonged to Florence Nightingale, T. S. Eliot, George Bernard Shaw, Princess Diana, President Ford, Jack Ruby (who killed Lee Harvey Oswald), the (converted) London gangster Reggie Kray, Queen Elizabeth II, Elvis Presley and others. The Museum of the Bible displays a microform Bible taken to the Moon on Apollo 14.

Bibles have been used in every age and Christian culture as charms, and it is sometimes for their apotropaic character rather than for their text or their bibliographic interest that they end up in museums (Malley 2013; Horsfield and Asamoah-Gyadu 2011). Of course, Bibles can be all three. The Book of Durrow, in Trinity College Dublin, is a magnificent seventh-century manuscript venerated for hundreds of years as a relic of St. Columba. Among its adventures was being dunked to create holy water to cure a farmer's cows. Quotations from the Bible's text can also be used as charms, and papers with quotations from the Bible are sometimes found hidden in old houses, stables and byres. They occasionally end up in museums, but these are rather seldom Bible museums, more often social history or folk museums.

Museums and the Art-lover's Bible

Sixteen years ago Ena Heller (2004: 124) drew attention not only to the way museums subordinate religion and religious objects to art, but how things were changing; they are still changing dramatically. There is an increasing literature on the role played by religion and religious objects in museums and on the role of museums in religion (Paine 2013; Buggeln, Paine and Plate 2017). More importantly though, museums are themselves more and more acknowledging the religious meanings of objects in their collections, recognizing the huge role religion plays in the cultures they are displaying and celebrating the artistic expression of human spirituality.

Ena Heller herself played a key role in that evolution as Director of the American Bible Society's late-lamented Museum of Biblical Art in New York. This was an exhibition gallery rather than a collection (though the Society itself has a significant Bible collection), which concentrated on exhibitions that explored, from an objective outsider's perspective, the territory between art and religion. Exhibitions were imaginative in their variety. The opening show in 2005 explored the work of self-taught artists from the American South and the influence on them of Evangelical Christianity. Other shows dedicated to modern artists followed, but there were also exhibitions on the uses of clay oil lamps through history, on loan from the Bible Lands Museum in Jerusalem, one on the King James Bible, one on 'Finding Comfort in Difficult Times: A Selection of Soldiers' Bibles' and another on Louis Tiffany stained glass. The last enormously successful exhibition, before the museum was obliged to close in 2015, displayed superb Renaissance works lent from Florence Cathedral.

Other Christian bodies use temporary exhibitions as part of their missionary effort and these often have a strong biblical theme. As the website of the Methodist Modern Art Collection puts it, 'Since the catacombs of Rome the gospel has been articulated as much in image as it has in word'. The Collection has some fifty works, many by well-known artists, and organizes touring exhibitions in the UK every year.[15]

Four Challenges for Bible Museums

This chapter has looked at a variety of types of Bible museum and briefly at the secular museums, exhibitions and libraries that also present the Bible. The examination has pointed up four issues, at least, that challenge many Bible museums today.

The first challenge particularly affects Bible museums that use archaeological finds to help tell the story of the Bible. Here the challenge is the same one that faces every curator in every museum worldwide: to use the objects to tell the truth as accurately as can be. In religious exhibits, though, 'what is truth?' becomes germane. Some Bible museums have been accused of misusing archaeological objects to support an understanding of the Bible's history that derives more from faith than from scientific evidence (Kutner 2018).

The second challenge is perhaps largely technical. Of their nature, museums are limited in the stories they can tell. Behind some Bible collections lies an urge to understand the text as the inspired Word of God and to understand it on a number of different levels. For many Bible scholars in the past, the Bible's text could be analyzed in four ways. The first and least important was the literal sense of the words: what the book says actually happened. The second was the allegorical, how (for example) Jonah's whale might symbolize the devil and its mouth might symbolize hell. The third level was the moral sense of the text, what it tells us about how we should live. Finally, the fourth and deepest level at which any biblical text could be profitably read was the 'anagogic', that is the mystical or spiritual sense of the text; 'it tells of one's soul and its relationship with God' (De Hamel 2001: 102). For centuries, examination of the Bible's text was to discover a range of meanings, not necessarily intended by the original author, but intended perhaps by God. For devout readers of the Bible today, it is perhaps the first and the fourth levels of analysis that seem most important; and for Christians there is a further dimension, the 'typological', by which almost every part of the Old Testament can be read as a prediction of the New.

On the one hand, the Bible can offer what has been called normative authority (Riches 2000: 50), and 'provide the means for the community to make decisions, to manage conflicts and to give rulings in matters of belief and practice'. But it can also offer formative authority and more subtly shape the attitudes, understandings and preoccupations of a society. Such interpretations still continue today. Many of the Bible's best-known stories have been invoked to answer contemporary questions and inform modern understandings.

Unsurprisingly, few Bible museums attempt to introduce the casual visitor to the subtle complexities of this tradition, key though it has been to the story of the Bible and its impact on such a variety of different societies. Rather seldom, too, do Bible museums examine the impact of the Bible on past communities. A few try to do so, more or less profoundly, most notably perhaps the Museum of the Bible in Washington, D.C. These questions are difficult to address in museums; museums can put the text on display, but cannot easily display the devout scholar and exegete, and his or her response to the text. Yet without trying to do so, is the museum truly presenting the Bible? 'If an interpretation concentrated only on what the biblical author said, and ignored the way generations of Jews and Christians had understood it, it distorted the significance of the Bible' (Armstrong 2007: 221). Will new technologies enable these issues to be introduced in a museum without boring its more casual visitors to death?

Bible museums do certainly take stories from the Bible and apply them to present-day questions, or use them as jumping-off points to address present-day issues. In 2015 The Jewish Museum in Berlin staged a big exhibition by Welsh film director Peter Greenaway and Dutch multimedia artist Saskia Boddeke, building on the story of Abraham's promised sacrifice of his son Isaac. Entitled 'Obedience', the exhibition asked 'Which is stronger, God's command or fatherly love? And where is the modern subject situated between the conflicting priorities of obedience and trust?' In 2016 the Witte de With Centre for Contemporary Art in Rotterdam, Netherlands, held an exhibition which addressed the eternal museum question: Do objects and images have fixed meanings, or are they inherently unstable and dependent on the meanings imputed to them by their users/observers? They called the exhibition 'In the Belly of the Whale' to reference Jonah's meditation there and the message with which he emerged. In 2018, the excellent Jewish Museum in Hohenems, Austria, based an exhibition on the role of borders in today's world on the story of the Gileadites and their password 'Shiboleth'; a series of installations examined borders, from the USA/Mexico frontier to the use of voice-analysis to determine asylum claims.

The third challenge facing many Bible museums is perhaps more profound. I was told at one Bible museum I visited:

> We try our best to share the love of Jesus Christ with all who visit, alongside the value of the Word of God. Our aim is to introduce those who have never read the Bible to it, and to encourage them to pick one up and read it.

That is the fundamental motive of almost every Bible museum. Most Bible museums are Protestant and evangelical (at least with a small 'e') in their ethos, though we have seen that a few Jewish Bible museums have an equivalent mission, and there are Catholic Bible museums, though comparatively rare.[16] Whether their second motive is love of collecting, love of art, love of archaeology or whatever, almost every founder of a Bible museum combines that with a love of the Bible as a text that offers salvation to the individual human soul. As one (anonymous) Bible museum curator put it: 'some Bible museums choose to focus on the revelation of who God is, how he works in the world and the significance of the life and death of Jesus'.

Here lies, perhaps, this third challenge to Bible museums. Every museum worthy of the name seeks to change lives; every museum worthy of the name does so by deploying well-chosen, well-managed and well-cared-for collections. Managing the balance between these two duties is the key role of the curator. Bible museums can sometimes find them particularly hard to balance because their mission to spread the Gospel can seem overwhelming.

It was this problem that lay behind the scandal that formed the background to the opening of the Museum of the Bible in November 2017. The museum's parent company, the craft-store chain Hobby Lobby, had recently been fined $3m for the illegal import of looted Iraqi antiquities, and there was evidence that a good many items in its collection[17] were fake, looted from archaeological sites, illegally exported or otherwise of very dubious provenance (Moss and Baden 2016; Davis 2017).

This illustrates one of two disturbing aspects of *collectors'* Bible museums. They come from a tradition of collecting where the main motive is the personal pleasure of ownership, with in many cases no recognition that other people and future generations may have a legitimate interest in their objects. As a consequence, detailed documentation, and research into where objects came from and what their history has been, is of little interest. In late 2017, however, Museum of the Bible placed on its website an admirable 'Provenance' page setting out the museum's acquisition policy and the results of detailed research into the first few key objects in the collection; this is being extended at least to everything on display.[18] Similarly the Van Kampen collection at the Scriptorium in Orlando was a few years ago in a chaotic state in its air-conditioned vault, with virtually no documentation and researchers largely excluded; recent reports suggest that things are improving fast.

The fourth challenge to Bible museums derives similarly from this collectors' tradition. It is that some Bible museums feel free to sell objects from their collection. For museum people, this is equally shocking. For us, the first duty of every museum is to preserve for generations to come both the object, and as much information about it as we can possibly retrieve; only so can a museum fulfill its mission and remain a true museum. For some Bible museums, though, the spreading of the Gospel is far more important. The Rev. David Smith[19] argues that the power of the Bible – of individual copies of the Bible and related items – to impact on individuals can be greater if they are owned by individuals who can actively take them around, show them to people and allow people to handle and engage with them; in museums things tend to get buried in store. He would be happy to see his collection sold and dispersed one day.

Such handling, though, will slowly but surely lead to their destruction, and sales bring a loss of information about each object that hugely reduces its value for personal engagement, not just for scholarship. Bible museums can be worthy of the name 'museum' without sacrificing their distinctive role and mission. We must hope that in the future every Bible museum will want to post the Code of Ethics of the International Council of Museums on its website, as Freiburg's *Bibel+Orient Museum* does today.

Conclusion

Despite these concerns, Bible museums have grown from small beginnings over the past century or so to become a worldwide phenomenon, explaining the Bible to countless visitors. Some have grown from the Bible collections of bibliophiles and these are often at the core of the larger

Bible museums. Archaeological collections underpin Bible museums that concentrate on telling the story of the Holy Land in 'the days of the Bible'. Bible museums can sometimes play an important political role, particularly in Israel, while – perhaps surprisingly – the impact that the Bible has had on so many societies is tackled by just a handful of Bible museums. The role the Bible as a physical object has played in the ritual of the synagogue, the church and the chapel is exemplified where historic ceremonial Bibles are displayed, and these too, we have argued, count as Bible museums. So too, sometimes, are museums which display works of art depicting scenes from the Bible, or inspired by its stories and message. But these are by no means the only reasons that Bibles appear in museums; they can be valued by curators for their associations and the uses to which they have been put in the past.

In some of these approaches it is the physical object that is to the fore, the book that has been venerated, displayed, elaborately decorated and which inspires by its form and presence. In others it is the book's story that the museum tells: the story of its creation and of its reception and use over two thousand years. In yet other museums it is the Bible's text and message that is presented to visitors. All of these different approaches can be found underlying Bible museums, but they are all Bible museums, and all make their contribution to explaining and celebrating the story and the message of one of humankind's most astonishing creations: the Bible.

Notes

1. The 'Materializing the Bible' website counts over seventy, but there are undoubtedly more in the generous definition of 'Bible Museum' used in this chapter.
2. James Watts (2013: 9) has proposed a threefold understanding of the Bible: as semantic (to do with meaning of the text), performative (performance of the words and performance of the contents, so including art and drama as well as reading and recitation) and iconic (every reverencing from its design to its liturgical procession). And see Horsfield and Asamoah-Gyadu 2011.
3. I am grateful to Richard Linenthal for talking to me so helpfully about the world of Bible collecting.
4. http://www.solagroup.org/vkc.html (accessed September 2016).
5. I am very grateful to Ellen Reid for this information.
6. I am very grateful to David Smith and Eva-Lotta Hannson for their welcome and kind help on my April 2018 visit.
7. I am very grateful to Paul Mathew and Binu Mathew for their welcome on my visit in 2015. See http://www.bsind.org/the_bible_world.html (accessed March 2018).

8. Though I fear she is probably still very discontented with this chapter, I am profoundly grateful for the careful critique Sarah Nightingale of Bibleworld offered of its first draft.
9. I am grateful to Dr. Eduard Baumann for these details.
10. http://www.blmj.org/en/template/default.aspx?PageId=2 (accessed March 2018). And see Boardman 2017. Mysteriously, this is one of two museums I approached who refuse to release visitor numbers.
11. Quoted in Roitman 2001: 60.
12. The museum now focuses exclusively on exhibitions, and has sadly dispersed its historic collection.
13. 'The Voyage of a Book' is an exhibition loaned by the Museum of the Bible.
14. I am grateful to colleagues from the Social History Curators Group for helpful comments and examples.
15. See http://www.methodist.org.uk/our-faith/reflecting-on-faith/the-methodist-modern-art-collection (accessed March 2018). And see Wollen 2000. Temporary exhibitions can play just the same role as actual Bible museums. There are touring Bible Exhibitions in a number of countries; my favorite is the 'Knitted Bible Exhibition' developed by the United Reformed Church in Hartlepool, northern England; it comprises knitted wool figures in 33 Bible scenes.
16. The Musée Bible et Terre Sainte was created in the Catholic Institute of Paris in 1969, and displays around 6,000 objects representing everyday life in Palestine from 5000 BCE to 600 CE; unfortunately, it is only open two hours a week during university terms. Museumpark Orientalis, a hugely attractive open-air museum outside Nijmegen in the Netherlands, began life in 1911 as an educational attraction beside a large Catholic pilgrimage center. Today it presents the three Abrahamic religions (Laarhoven 2014). While there are few museums devoted principally to the Torah, most Jewish museums have a Judaism/Judaica section, and often a school program centered on religious practice.
17. The 'Green Collection', legally owned by Hobby Lobby, is said to comprise 40,000 objects; the 'Museum Collection' comprises 2,559 objects, some 1,159 of which are on display.
18. David Trobisch, pers. com., 16 April 2018.
19. The Rev. David Smith, pers. com., 27 April 2018.

Bibliography

Ambrose, T. and C. Paine (2018), *Museum Basics: The International Handbook*, 4th edn., Abingdon: Routledge.

Anon. (2018), 'Abba Garima Monastery: A Monastery's Treasures Are the Focus of a Row over Heritage and Conservation', *The Economist*, 426: 9084.

Aronsson, P. and G. Elgenius (2015), *National Museums and Nation-Building in Europe 1750–2010: Mobilization and Legitimacy, Continuity and Change*, Abingdon: Routledge.

Armstrong, K (2007), *The Bible: The Biography*, London: Atlantic Books.

Boardman, J. (2017), *Dr. Elie Borowski: Founder of the Bible Lands Museum Jerusalem*, Jerusalem: Bible Lands Museum.

Buggeln, G., C. Paine and S. B. Plate, eds (2017), *Religion in Museums: Global and Multidisciplinary Perspectives*, London: Bloomsbury.

Carroll, S. (2007), 'Biblical Treasures in Private Holdings: The Van Kampen Collection', in J. Wineland (ed.), *The Light of Discovery: Studies in Honor of Edwin M. Yamauchi*, The Evangelical Theological Society Monograph Series 6, 235–93, Eugene, OR: Pickwick Publications.

Davis, K. (2017), 'Caves of Dispute: Patterns of Correspondence and Suspicion in the Post-2002 "Dead Sea Scroll" Fragments', *Dead Sea Discoveries*, 24 (2): 229–70.

Ham, K. (2008), *Journey Through the Creation Museum: Prepare to Believe*, Green Forest: Master Books/Answers in Genesis.

De Hamel, C. (2001), *The Book: A History of the Bible*, London: Phaidon Press.

De Hamel, C. (2016), *Meetings with Remarkable Manuscripts*, London: Penguin Books.

Heller, E. (2004), 'Religion on a Pedestal: Exhibiting Sacred Art', in E. Heller (ed.), *Reluctant Partners: Art and Religion in Dialogue*, 122–41, New York: The Gallery at the American Bible Society.

Horsfield, P. and K. Asamoah-Gyadu (2011), 'What is it about the Book? Semantic and Material Dimensions of the Word of God', *Studies in World Christianity*, 17 (2): 175–93.

Knell, S. et al. (2011), *National Museums: New Studies from Around the World*, Abingdon: Routledge.

Kutner, M. B. (2018), *Hobby Lobby's Museum of the Bible Steals; Does it Also Lie?* Eidolon, https://eidolon.pub/hobby-lobbys-museum-of-the-bible-steals-does-it-also-lie-ee09a3335e3f (accessed October 2018).

Laarhoven, J. van (2014), 'Museumpark Orientalis, Nijmegen, Netherlands', *Material Religion: The Journal of Objects, Art and Belief*, 10 (2): 252–54.

Lampe, G. W. H. (1969), 'The Exposition and Exegesis of Scripture', in G. W. H. Lampe (ed.), *The Cambridge History of the Bible, Volume 2: The West from the Fathers to the Reformation*, 155–279, Cambridge: Cambridge University Press.

Malley, B. (2013), 'The Bible in British Folklore', in J. Watts (ed.), *Iconic Books and Texts*, 315–47. Sheffield: Equinox.

van der Meer, M. N. (2010), 'The Biblical Museum Amsterdam', *Material Religion: The Journal of Objects, Art and Belief*, 6 (1): 132–5.

Moss, C. R. and J. S. Baden (2016), *Bible Nation: The United States of Hobby Lobby*, Princeton: Princeton University Press.

Paine, C. (2013), *Religious Objects in Museums: Private Lives and Public Duties*, London: Bloomsbury.

Paine, C. (2019), *Gods and Rollercoasters: Religion in Theme Parks Worldwide*, London: Bloomsbury.

Rakow, K. (2017), 'The Bible in the Digital Age: Negotiating the Limits of "Bibleness"', in M. Opas and A. Haaplalainen (eds), *Christianity and the Limits of Materiality*, Bloomsbury Studies in Material Religion, 101–21, London: Bloomsbury Academic.

Riches, J. (2000), *The Bible: A Very Short Introduction*, Oxford: Oxford University Press.

Roitman, A. (2001), 'Exhibiting the Dead Sea Scrolls: Some Historical and Theoretical Considerations', in N. Silberman and E. Frerichs (eds), *Archaeology and Society in the 21st Century: The Dead Sea Scrolls and Other Case Studies*, 41–66, Jerusalem: Israel Exploration Society.

Roitman, A. (2006), *The Bible in the Shrine of the Book: From the Dead Sea Scrolls to the Aleppo Codex*, Jerusalem: The Israel Museum.
Schefzyk, J. (2016), 'Bibelmuseen', in M. Klöcker and U. Tworuschka (eds), *Handbuch der Religionen: Kirchen und andere Glaubensgemeinschaften in Deutschland und im deutschsprachigen Raum*, 50, Bamberg: Mediengruppe Oberfranken.
Sturm, T. (2017), 'Christian Zionism as Religious Nationalism Par Excellence', *Brown Journal of World Affairs*, 24 (1): 7–21.
Summers, C. (2017), *Lifting Up the Bible: The Story Behind Museum of the Bible*, Washington, D.C.: Museum of the Bible.
Taylor, E. P. (2015), 'Producing the Dead Sea Scrolls: (Trans)national Heritage and the Politics of Popular Representation', MA thesis, University of Massachusetts Amherst.
Watts, J. (2013), 'The Three Dimensions of Scripture', in J. Watts (ed.), *Iconic Books and Texts*, 9–32, Sheffield: Equinox.
Wharton, A. J. (2006), *Selling Jerusalem: Relics, Replicas, Theme Parks*, Chicago and London: University of Chicago Press.
Wollen, R. (2000), *The Methodist Church Collection of Modern Christian Art: An Introduction*, Trustees of the Collection.
Wright, C. E. (1957), *Biblical Archaeology*, Philadelphia: Westminster Press.
Yoder, D. (2016), *The German Bible in America: An Exploration of the Religious and Cultural Legacy of the First European-Language Bible Printed in America*, Kutztown: Pennsylvania German Cultural Heritage Center.

Chapter 9

REWRITING THE BIBLE:
THE VISUAL CULTURE OF CREATION SCIENCE

Larissa Carneiro

In one of the many exhibit rooms at one of the most significant creationist museums in the world, a young boy did not look very amused by a series of exhibits displaying biblical scenes. 'I want to see dinosaurs!' he cried out. His mother shushed the boy and warned him that he should pay attention to everything the museum had to teach, biblical scenes included. The dinosaurs would have to wait.

The scene I witnessed was telling. As the mother advised her impatient son, the Creation Museum is not a place only to have fun. Owned and operated by the apologetic ministry *Answers in Genesis*, the famous tourist destination in Kentucky is committed to two very important missions. First, to argue against the premises of two overarching scientific paradigms – Charles Darwin's evolutionary theory and Charles Lyell's idea of uniformitarianism, the view that physical changes occur gradually over long periods of time; second, to contend for the scientific plausibility of another paradigm known as Young-Earth Creationism, Creation Science, or Flood Geology. In pursuit of the second task, the museum wants to persuade viewers that everything described in the first chapters of the Bible literally happened. For Young-Earth creationists, the book of Genesis is more than a collection of ancient folk stories. It is an accurate scientific and historical account, which means that, according to the creationist perspective: the universe and humankind were created in six literal days by nothing more than divine utterance; all humans are direct descendants of the first couple, Adam and Eve; people co-existed with dinosaurs; the Earth is no older than 10,000 years; and all major geological formations were either caused by the act of the divine Creation itself or, 1,656 years later, by a cataclysm of global proportions: the biblical Flood related in the story of Noah.[1]

In order to provide what creationists consider as scientific evidence against evolution and uniformitarianism, the Creation Museum takes a shape that is not very different from its secular counterparts. The place mimics the visual and rhetorical strategies of natural history museums: visitors encounter an array of dioramas, charts, graphics, videos, replicas of dinosaurs, displays of fossils and all sorts of material artifacts. Even the famous Lucy, categorized by mainstream science as the last common ancestor of chimpanzees and humans – and therefore a mandatory presence in every natural history museum – is there, although she is not represented as a distant relative, but as a mere ape.

But I will not dwell on the similarities between secular scientific institutions and the Creation Museum.[2] My focus here rests on what is *not* to be found in any of its secular counterparts: the representation of religious scenes. While this unique trait makes the museum ludicrous for those who do not subscribe to the inerrancy of the Bible, neither ridicule nor parody are my interest. I want instead to explore the role of visual culture in servicing the creationist understanding of the relationship between science and religion. By 'visual culture' I mean not only the imagery that creationists use, but the forms of display, the visual fields for engaging the imagery and how images assemble relationships to texts and discourses. Visual culture, as David Morgan has argued, is best understood as ways of seeing that bring viewers into interaction with images (Morgan 2012). I will show how the Creation Museum visually engages visitors in re-imagining the Bible, doing so in a way that provides what is not literally in the text.

Science and Religion

The Creation Museum is the materialization of what Stephen Jay Gould called an 'oxymoron' in science (Gould 1997: 8). In its vast building, there is no such a thing as 'nonoverlapping magisteria', the clear separation of what Gould considers two very distinct domains: science and religion. 'Science is the empirical constitution of the universe', while religion is in the business of searching 'proper ethical values and the spiritual meaning of our lives' (9). Contrarily, the Creation Museum dares to be a place where scientific and religious imagery intermingle and share the same space. It proudly mixes two areas that in the realm of so-called secular science should not be mixed. In fact, representations of biblical scenes dominate the museum's permanent installations. If creationists do not invest considerable financial resources in actual research, they certainly spend substantially in order to demonstrate in the material terms of exhibits what they consider is depicted in the book of Genesis. The museum's motto, 'Prepare to Believe', presumes the popular sentiment that seeing is believing. There is evidence to be found in the Bible, creationists insist, that authorizes the step from one to the other, from seeing to believing. Indeed, there is even the assumption that because we see it, it must be true.

To prove that the events described in the book of Genesis really happened is so fundamental for the premises of Young-Earth Creationism, that in 2016 the evangelical and fundamentalist project *Answers in Genesis* opened to the public an even more ambitious project than the Creation Museum. The Ark Encounter joined the museum in the Christian mission to bring the pages of the Bible to life. Only 44 miles away from its sister attraction, the Ark Encounter can be described as an immense biblical setting dedicated exclusively to Noah's story. It features a full-size replica of the Ark measuring 510 feet-long, 85 feet-wide and 51 feet-high. The first of three decks contains 132 bays displaying models of all kinds of animals that creationists believed were carried in the vessel. On the second deck, viewers encounter more animals and also dioramas portraying Noah and his family involved in daily activities during their epic journey. Finally, the third deck shows scenes that creationists believe might have happened inside and outside the Ark during the Flood.

The energy and ambition of the Creation Museum and the Ark Encounter should not conceal the significant problems in (re)producing factually what is described in the Scriptures. This difficulty resides precisely in the content of the book of Genesis. Even the most careful readers will not find details about daily life inside the Ark. Genesis also does not specify how Noah could conceive and develop such a complex project. Neither does it describe the sort of knowledge, materials and tools that Noah and his family employed to construct and to operate the Ark. Consequently, what happens to the Bible when scenes that are only partly described are added to the narrative? How will the book of Genesis come to life, as promised by the Creation Museum and the Ark Encounter? What kind of religious imagery emerges when a full-scale representation of 'the actual ark' aims to demonstrate accurately what kind of materials, tools and skills were used to construct the vessel? Or when these representations meticulously describe rooms and artifacts supposedly extracted from the biblical narrative? In fact, the high degree of conjecture that the

Ark's creators had to engage in during their design phase, owing to the slender descriptions and the brevity of the entire account in the Bible, is not something they tried to avoid, but instead regarded as a creative challenge.[3]

But this very effort can be shown to incorporate a modern scientific sensibility. Creationists are not immune to its effect. Visual analysis of scenes of Noah's office in the Creation Museum and in the Ark Encounter will highlight the creationist application of scientific thinking: creationists want to explain as much as possible the plausibility of the Ark's construction and success rather than enshroud it in a cloud of miracles. The choice of the subject is not random. It is Noah himself and the construction of the Ark that require more free interpretation, more speculation and consequently, more work to fill the gaps in the biblical narrative. It will become clear that the problem posed by the Creation Museum and the Ark Encounter is not just a scientific one – it is also religious. Contradicting their supposed commitment to 'biblical inerrancy', creationists have to cross the boundaries of the Bible's literal content. The simple fact is that most of the information they require to design and build a full-scale model literally is not to be found in the Bible. They must infer it or otherwise supply it by creative problem-solving. In order to engage the religious book as a scientific enterprise, the museum and theme park work as milieus for resolving ambiguities, gaps and anomalies presented in the Bible. In short, creationists must re-write the book of Genesis in order to prove that the construction of the Ark was feasible in biblical times.

I turn now to a suggestive essay by Bruno Latour since it helps us discuss the nature and functionality of religious and scientific imagery. Then I will undertake a visual analysis of historical Protestant and creationist images of Noah and the Ark to demonstrate how Young-Earth Creationism, in spite of defending the inerrancy of the Bible, is part of a long Protestant tradition of re-interpreting and re-writing the Bible according to modern and scientific sensibility.

'Person-making' Images vs. the Hermeneutics of Literalism

An image always comes with a point of view, a way of seeing, a positioning of the viewer in relation to a subject and to other viewers. Michelangelo's portrayal of the story of Noah, for example, painted on the ceiling of the Sistine Chapel (see Fig. 9.1), tells the story of the Great Flood from the perspective of those who are doomed, desperately struggling to save their lives from the flood, but fated to die.

9. CARNEIRO *Rewriting the Bible* 205

Figure 9.1. Noah and the Ark, Sistine Chapel, Michelangelo, 1508–12, fresco. Image courtesy of Wikipedia in public domain.

The image shows that God's punishment makes no discernment of gender or age. On a piece of land that had not yet been taken by the water, a mother tries to cover and protect her baby, cradling it in her arms, while a toddler holds her leg, seeking for maternal protection that is fleeting at best. Not far away, an old father carries the corpse of his dead son. There is no hope for them either. A family carries its few belongings to a last bit of dry land. We know that the act of saving material possessions will soon be worthless because everyone in this scene will shortly perish, including the horse whose head is visible on the far left. The mythical Ark, a square wooden box, the only thing that will survive the horrible cataclysm, is not in the foreground of the painting. On the contrary, the vessel is visible only in the distance – a small island of godly safety amidst the tragedy. An improvised boat tries to reach it. A group of people has already got to the Ark, eager to be spared from divine wrath. A man with an ax tries to break in, forcing his entry and pressing his salvation. Other men use a ladder to get to the window at the top of the vessel. But it is pointless. The Ark is already sealed against all living creatures and humankind that ignored God's commandments. On one of the Ark's sides, we see the only man whose family was spared, the man who was ordered to save one pair of all kinds of animals in the world. Emerging from a window in the craft, a bearded Noah seems to be interacting with the divine, oblivious to the people's tumult and despair. At the top of the ark, a white dove refers both to the bird that was later sent to find dry land and also to the power of salvation that only the Holy Ghost can offer. And that is all we can see of the Ark – no other animals, no interior and no other family members.

Compare Michelangelo's scene with another depiction of Noah, which I will describe. This representation is not a painting, but a three-dimensional diorama with an animatronic Noah designed by Doug Anderson, installed in 2009 in the Creation Museum. A life-sized Noah sits before the visitor, an iconic figure featuring a long, gray beard, wrinkled eyes and a long-sleeved, mustard-colored tunic. We do not see the outside world, the fury of the water, the cloudy sky, nor the despair of those who did not escape divine wrath. The diorama focuses our attention solely on Noah and the interior of the Ark. Sitting in his private studio, the aged Noah works at a well-crafted writing desk, which has carved on its side scenes of the six days of Creation. Where Michelangelo relegated the major character of the story to the back of his composition, the Creation Museum scene foregrounds Noah. Immersed in his own thoughts, he holds a pen and seems to be writing a day-by-day log of his journey, using what appears to be the Babylonian numeric system and cuneiform characters. On the top of his desk, there are other folded documents apparently made from pulp rather than papyrus. On the shelves in front of him are a series of pots, probably containing ink, and more pens ready for use. Beside the paper, a star chart made of clay lies on the desk. If asked (by using a touchscreen monitor), the animatronic Noah comes to life to answer practical questions about the Ark, such as how he could accommodate so many animals inside it, whether or not it contained dinosaurs and how they could fit despite their size.

For some assistance in thinking about how these images operate within the domains of science and religion, I turn to a fascinating discussion by Bruno Latour (1996). In 'How to Be Iconophilic in Art, Science, and Religion?', he examines what he considers the similarities and the differences between religious and scientific representations. First, they are both highly mediated, since information, regardless whether scientific or religious, is never simply transferred, but always 'radically transformed from one medium to the next'. Latour describes such transformations as 'iconophilia', an attitude, he stresses, that is not 'respect for the image itself but for the movement of the image', for what the image does (421). Both in science and religion, this respect is expressed through the requirement of giving a visual form to ('in-form') either a natural phenomenon or a moral principle. In science, iconophilia manifests as the imperative to transform the crudeness and unpredictability of nature into stable scientific results: for instance, the invisible atom is transformed into a visible drawing and the ungraspable natural landscape into a portable map. In religious images, the process of 'in-formation' occurs when the narrative of the Bible is transformed into visual representations: a religious painting on the ceiling of the Sistine Chapel, for example. Still regarding their similarities, both forms of representation aims to provide access to what is in fact invisible to naked eyes, to the inaccessible, to the a-historical, to what is represented to look like 'an unmediated essence', although absent (427).

But in what amounts to Latour's own version of 'non-overlapping magisteria', the similarities end here because the production of scientific and religious images is driven by different ends. In science, the transformation of a landscape into a map occurs for the sake of disseminating portable, reliable and immutable information. A map is not the territory it describes: it is impossible to carry the entire landscape to find a location in it. But it is possible to find one's way through the territory if the landscape is 'trans-formed' into printable materials that can be read and carried by anyone anywhere (424). In science, Latour adds, nature needs to be transformed into different forms of inscriptions (drawings, charts and numbers) in order to be disseminated across time and space (425).

However, according to his argument, in religion the process of in-forming the Bible into visual representation has a different objective: religious images have the moral purpose of individual formation, or what Latour calls 'person-making' imagery (428). He considers it a mistake to think that religious representations are in the business of literally re-presenting the 'facts' described in the book of Genesis or in the Gospels (430). In contrast to scientific pictures, effective religious images do not aim to provide access to the natural or the historical world. For him, this is the realm – or the magisterium – of science. The objective of religious images is to warn individuals to behave properly, to pray, love, meditate, fear and avoid the path of a sinful, non-Christian life (431). Religious representations, he continues, are not about a scene from the New or Old Testament. The core of his argument is this: unlike images found in scientific books or natural science museums around the world, religious scenes are not in the business of accurately representing or explaining the past. They represent what they *mean* – meanings that may be temporarily forgotten, but are constantly recalled and re-understood in the act of seeing devotional pictures. Religious images are in the business of reminding viewers of the fundaments of their faith. They do not refer to the past but renew the present. It is not about what happened then so much as about how viewers should behave now for the sake of their souls.

Michelangelo's painting in the Sistine Chapel might be taken to support Latour's argument. The artist's portrayal of the Great Flood does not stand or fall on literally representing what the book of Genesis describes. The Flood, the Ark and Noah are there but the painting's tragic depiction of divine wrath seeks to focus attention on the consequences of human disobedience and divine punishment. Michelangelo was not transforming factual information about the past into a form of visual information. Guided by his sponsors, his intention was to warn about what happens when people choose a wicked path instead of a Christian one. Michelangelo did not intend to describe the ancient world as it was, but to dispose viewers to respond to the image's point of view. Looking at the painting looming far above the chapel's floor, viewers are not urged to ponder literalisms, factual details, the specific landscape, or the genuine event in the scientific sense. The image invites them, Latour would argue, to focus on what it means to be and act as Christians in the present.

But what about the depiction of Noah in the Creation Museum, a place that claims to be a natural science museum in which scientific and religious representations work together? What is the purpose of religious imagery that claims to represent the past in a historical and scientific way? Like Gould, Latour dismisses Creationism out of hand as a mere 'hilarious attempt'. 'Creationists are an excellent demonstration that some Christians can be rationalized to the marrow, unable even to retrieve a shred of the kind of talk that would not carry information but transport persons' (434). Latour may correctly describe creationists as 'rationalized to the marrow', but his claims miss the boat on two counts. First, he builds up his argument on the differences between religious and scientific representations by analyzing only Renaissance examples of Catholic visual piety. In the nineteenth century, for instance, Catholic painters became very interested in documenting and portraying the historicity of the life of Christ. Some travelled to the Holy Land to study its people, landscape and architecture as the basis for their religious art.[4] Before that, in the seventeenth century, the Jesuit Athanasius Kircher tried to reconcile the events described in Genesis with new scientific knowledge. Media theorist Friedrich Kittler (2010) even asserted that Kircher envisioned the *camera obscura* as a scientific device for creating the illusion of reality of biblical scenes.[5] Second, Latour scorns the creationist use of images by contrasting them to the Catholic tradition, ignoring a whole history of Protestant imagery. The art historian David Morgan has written extensively about the history of Protestant visual piety, arguing against the common notion that Protestants do not use images in their religious practices.[6] They do, but differently from Catholics. For Morgan (2007), a dominant function of images in Protestantism has been to underscore the 'iconicity of the text', which means that the Bible becomes a conduit of undistorted truth (222).

Rather than simply a 'hilarious attempt', the Creation Museum is part of a larger Protestant visual tradition in which the Bible is perceived to be the correct guideline for understanding history and the Book of Nature, the rational world of natural laws. Consequently, in the galleries of the museum, representations of scenes of the book of Genesis do not intend only to teach viewers how to behave but, similar to natural history museums, how the past actually happened.[7] Dioramas representing Noah and his family are not ominous sermons like Michelangelo's painting, but creationist versions of full-size tableaux representing *Homo erectus*, *Homo sapiens*, Neanderthals and Cro-Magnons, just as we find them, for example, in the Hall of Human Origins at the American Museum of Natural History.

Consequently, it is not ludicrous to state that Young-Earth Creationism imagery has its foundation in the rise of modern science. It is evidence of how much scientific sensibility has affected and changed religious views over the past several centuries. From its beginning, Protestantism shaped and was dialectically shaped by emerging scientific discourses, which continued down to the early twentieth century in the United States, culminating in Creation Science. Fascinated by the scientific enterprise, Protestants intensified a process of interpreting the Bible as a factual book that could, like nature, be measured, analyzed and submitted to scientific scrutiny. Therefore, Protestant visual culture reflects this modern sensibility. Images, Morgan (2007) explains, became a device to endorse the transparency of text (222).

Yet by establishing a cult of the written word, Protestants have paradoxically re-scripted the Bible in the name a putative 'literalism' or 'inerrancy'. Biblical institutions, such as the Creation Museum, build their visual constructions based on bits of biblical literature, some scientific artifacts and much creativity and speculation.

The Bible as a Historical Book and Noah as a Man of Science

Genesis only provides a few clues about the Ark. It states that God ordered Noah to make himself a vessel of gopher wood. God also specified that the Ark should have sealed rooms: 'Rooms shalt thou make in the ark, and shalt pitch it with and without with pitch' (Gen. 6.14). Additionally, God stipulated the vessel's size. 'The length of the ark shall be three hundred cubits, the breadth of it fifty cubits, and the height of it thirty cubits' (Gen. 6.15). The size sounds impressive, though the proportions fail to describe a seaworthy craft. Yet the biblical author added further details to complete the reader's mental image. God told Noah that the Ark should have a window finished to a cubit from the top and a door was to be set in the side. The vessel should also contain three floors (Gen. 6.16). At the end of this narrative, it is said that Noah did exactly what God instructed (Gen. 6.22).

In spite of the lack of more detailed information, in the seventeenth century, Dutch publishers Hermannus Ribbius and Anthony Schouten engaged Constantin de Groot, a 'lover of Jewish antiquities', to write a comprehensive study of the ancient biblical world in which biblical narratives would receive the historical interpretation they deserved. For them, it was an important task since, like many Protestants, they considered it a mistake to represent (and distort) biblical narratives as mere religious allegories. In their view, biblical narratives such as the deluge had been historically misrepresented by painters who created religious images informed solely by their own imagination and poetic fables. This impaired the achievement of a rational and accurate comprehension of the Bible. For these Protestants, it was a mischief that drew upon the grossest superstition and illusions represented by the Catholic Church. In their two-volume book first published in 1690, they devoted themselves to correcting the fault. *Voor-Bereidselen tot de Bybelsche Wysheid* (Preparations unto Biblical Wisdom) was intended as an encyclopedic study of the ancient Jewish and Near Eastern world that would prepare the reader for the proper interpretation of scripture.[8] Across its 1,046 pages, historical and biblical characters and events were mixed together. Eighty-seven engraved plates and thirty-seven in-text engravings depict both biblical scenes and historical artifacts, such as historical coins, vestments, scrolls and documents, writing tools, maps and astronomical observations, among others. Historical events, characters and artifacts lend to biblical narratives' legitimacy. The project represents an important shift in how biblical passages were represented. The engravings by Wilhelmus Goeree did not intend to portray religious scenes as 'person-making' images, as described by Latour, but as historical ones. The third chapter of volume 2, for instance, is dedicated to the history of Noah and his Ark. Here, text and image describe and explain the event within a technical and scientific perspective. Instead of a horde of people trying to escape the inexorable fate of the rising water, we encounter the exposition of rational explanations for the feasibility and functionality of the vessel. How could Noah, an old man who lived in a desert region 4,000 years ago, construct a massive ship that was not only able to float above the troubled waters, but was also to survive a cataclysm of global proportions? What kind of knowledge was necessary to undertake such an enterprise? What kind of tools and materials did he use? In an illustration on page 213, we learn that he did not build the Ark with only his family, but that he probably hired people to help him. In this scene, cranes, ramps and diverse tools are employed in what is represented as a massive operation.

A host of other issues confronted Constantin de Groot and the illustrators, especially those concerned with natural laws. How deeply could the Ark go under the water without sinking? What cargo could it load and carry safely? An illustration answers some of these questions. It represents how Noah, guided by God, employed scientific knowledge to develop his enterprise (see Fig. 9.2). The image is divided in three horizontal sections. In the first section, three putti conduct an empirical experiment to demonstrate what looks like Archimedes' principle of buoyancy. Archimedes' principle is one of the natural laws of physics, fundamental to the mechanics of fluids and, consequently, to the development of navigation. This first segment works as a sequence of events. On the left, the first putto dropped a wooden box in the water, which now lies at the bottom of the tank. In the middle, the second putto uses a plumb level to measure the vertical axis of the box that is beginning to surface, acted upon by an upward force, or buoyancy. Finally, the third figure employs a scale to measure the displacement of the water and the weight of the box that now is almost floating.

Figure 9.2. Wilhelmus Goeree, engraver (1690). Voor-Bereidselen Tot de Bybelsche Wysheid, p. 248, vol. 2. Collection of the author.

In the second section, the seventeen-century illustration connects the ancient narrative with modern enterprise. It reminds the viewer that the basic knowledge of this natural law also made possible the flourishing of navigation and mercantilism, a polity of commerce dominant in the Netherlands, where the image was produced. Launched with the assistance of a group of putti, the ship is the workhorse of international trade, a leading force of which was the Dutch East India Company. At the bottom, we see the Ark, designed according to the information provided by the book of Genesis, but now represented not only as a plausible initiative, but also as the archetype of modern scientific and commercial enterprise. In classic illustrations, putti work as illustrative devices to demonstrate divine operation. But in the modern world, if the earth was still supposed to function according to God's power, it was no longer a place for endless miracles: events obeyed the natural laws of nature created by Him.

This motif of the Ark represented not as a religious allegory but as historical and technical fact is recurrent in Protestant illustrations. For instance, in another set of illustrations in a biblical dictionary (Woord-Boek) from the early eighteenth century, a triptych describes in more detail the features that an efficient Ark would have (Calmet 1725–31). The first image unfolds as a panoramic view of the enormous three-floor rectangular vessel. Dry land and lush vegetation are visible, suggesting that the waters are already subsiding. In spite of its three-floors, the Ark depicted in this picture does not resemble too closely the one described in the book of Genesis. From one single window on the top, the image shows the second and third floors as long sequences of thirty large windows, which logically would make the air more breathable and consequently better for the health of people and animals inside. On the second floor, appearing in each window, two animals peek at the world outside. In this image, which goes beyond the details provided by the biblical narrative, we are not only informed that the Ark carried one pair of all kinds of animals, but also how they were lodged. On the third floor, a person (probably Noah) lifts the shades of one of the windows to check if the time has come to unload its precious cargo: those who will repopulate the earth. In an inset above, a thumbnail also illustrates something that is not in the Scriptures: Noah engaged in the daily activity of feeding and taking care of the animals.

Also interesting for the argument of this chapter are the second and third parts of the triptych composed of architectural drawings of the Ark. In one, whose titles are 'Face and Appearance of Noah's Ark' and 'Cross-section of Noah's Ark', we are introduced to two different technical images of the Ark. At the top, the first shows a frontal elevation, which is a full view of the Ark seen from one side. At the bottom, the illustration displays a cross section of the Ark, which represents a vertical plane cut through the vessel, making it possible to have a clear view of its interior structure and, therefore, of how the ark was internally organized. According to the drawing, the bottom floor was composed of large cellars and was probably used for storage of food and other supplies; the second served to contain the animals, a large stable with individual stalls; and, finally the third floor was dedicated to Noah and his family. To enhance the level of technical information, at the bottom of the illustration a scale gives the viewer an accurate perspective of the size of the Ark.

Finally, the third part of the triptych, entitled 'Sketches of the Three Floors', illustrates the vessel's floor plan. A floor plan is considered the most fundamental architectural diagram. A view from above, the technical representation shows the arrangement of spaces in the Ark in the same way as a map. Architectural drawings are made according to a set of technical conventions, such as units of measurements and scales. Among their purposes, they are used (1) to assist builders in their enterprise and (2) as a record of the completed work, a subject that I will discuss shortly. For now, regarding the first purpose, the illustrations described above imply that the Ark did not simply appear by the sheer will or utterance of God's word. The Ark was not entirely a miracle. The project was carefully planned and its execution involved scientific and technical knowledge. Moreover, the complexity of the drawings suggest that the construction and navigation of the vessel could not be a job conducted by an ordinary man chosen for the divine mission based only on his moral character and devotion to God. It was directed by a man highly trained in technical and scientific skills.

More than three centuries separate these illustrations from the biblical settings displayed in the Creation Museum and in the Ark Encounter. When the Dutch illustrations were conceptualized, neither evolutionary theory nor uniformitarianism had entered the scene. During the second half of the nineteenth century, the Western academy underwent a paradigmatic revolution that toppled the idea that the universe and humankind were created by a supernatural act and that geological formations were caused by a series of cataclysms, such as the mythical Flood. However, even if this genre of religious images no longer held the status of historical visual accounts for a major audience, they indeed contributed to establishing the style and function of Protestant imagery: not 'person-making' tools, but devices to affirm the transparency and inerrancy of the biblical text.

Therefore, following a Protestant visual tradition, the representations of Noah's office exhibited in the Creation Museum and the Ark Project replicate features that are present in the seventeenth-century illustrations described above. First, the settings depict biblical narratives alongside historical artifacts in order to provide historical and scientific legitimacy to religious texts. Scrolls and documents, writing tools, maps and devices for astronomical observations are mandatory objects in both sites. Secondly, the representations are committed to the mission to create a background to explain that Noah was capable of building the Ark and performing the many duties required of him. What sort of career did he have in order to be chosen to perform such a complex task? As posed by the Ark Encounter website, '"Did he have multiple occupations prior to building the Ark, after all, he lived five centuries before his work on the Ark?' Did he have technologies to assist him in such difficult task? Did he know about shipping and civil engineering?[9] As stated in the website of *Answers in Genesis*, 'using a bit of imagination' it is possible to cogitate the idea of Noah as a very educated and skilled man.[10] Finally, both sites go beyond the modest description of the Ark offered by the book of Genesis and represent how the 'actual' Ark should have been in order to be functional.

So let's go back to Noah's office, first in the Creation Museum and then in the Ark Encounter. The objects on Noah's desk reveal many aspects of his personality that are not described in the Scriptures. Here we encounter a man represented as a prototype of the ideal modern man, guided by reason and trained in various technical skills. According to the Creation Museum, Noah had probably spent part of his days working in his office. By looking at him, we immediately learn that Noah could read and write. With a pen in his hand, immersed in his own work, Noah busies himself at his desk. What is he doing? Was he recording the geological event he had just witnessed? Did he describe how dark clouds had obscured the sun and how the rain had started to pour violently over the world? Was he writing a detailed report about the daily activities in the Ark and the unprecedented events that would have befallen his family during that year-long ordeal? If this setting in the Creation Museum wants to encourage the viewer to imagine the content of long-vanished documents, its motif goes beyond the attempt to evoke individual exercises of imagination. It intends to represent Noah as a technician, an architect, an engineer and a recorder of the pre-Flood time. Beside the paper before him appears a proto-astrolabe made of clay. The circular artifact bears an astonishing resemblance to a Neo-Assyrian tablet with depictions of constellations currently in the collection of the British Museum. The celestial planisphere is believed to be an instrument for astrological calculations representing the night sky over Nineveh on 3–4 January 650 BCE.[11] Does the Creation Museum facsimile imply that Noah had recorded the skies during the event of the deluge? In so many ways, the Museum's Noah is very different from the man simply interacting with the divine in Michelangelo's painting. The setting shows a man trained in mathematics, and one who commanded nautical skills by which he might orient himself based on the movement of celestial bodies, as determined by the use of astrolabes. And in a move that implies the historicity of the biblical account, Noah appears as a historian of his own time, chronicling an event that was no myth, but a historical occurrence.

The project for imagining Noah as 'jack of all trades' does not end in the Creation Museum. It is even enlarged in the Ark Encounter project. For the full-size ark, the designer Allen Greene conceptualized a much more spacious and sophisticated studio for Noah. The crew involved in designing the Ark aimed at constructing more than the vessel modestly described in the Bible: similar to the architectural engravings from the seventeenth century, they represented in detail the interior of the three-deck ship. In doing so, the creationist team sought to provide answers to prosaic questions that cannot be found in the book of Genesis. For instance, after the animals were fed and the cages were cleaned, what did Noah and his family do on the Ark? We learn that by encountering Noah in his office, working in the devoted company of his wife. The setting suggests that the couple spent their leisure time in intellectual and artistic activities in their private office. His wife, who remained unnamed in the book of Genesis, is also represented as a person of letters. Standing up, wearing a linen dress and jewelry, she reads what could be one of the documents that her husband probably wrote or maybe, as suggested in a post on the Ark Encounter blog, a letter from her now deceased mother.[12] Visitors can see brushes for painting, a sketch of a portrait, papyrus rolls, sheets of vellum, writing utensils, cylinder rolls of cuneiform text and a writing desk at which Noah – looking much younger than in the Creation Museum – is at work.

The conjectural work of the exhibit's creators is evident in the artifacts in Noah's studio, many of which come from different times in the history of the ancient Mediterranean. The Neo-Assyrian disc of clay is also there and hanging on the back wall. We can see two paintings of different men. Are they two of Noah's three sons? His ancestors? Regardless of whom these paintings represent, the style and format of these paintings unmistakably resemble mummy portraits of Roman Egypt, consisting of a practice and technique were traditionally used around 100–300 CE (Walker 2000). The two paintings bear a striking resemblance to the actual portrait of a young man produced around 125–50 CE and painted in encaustic on wood.[13] The likeness of the two images suggests that this image (produced at least 3,000 years after the date creationists established for the Flood) worked as a model for designer Allen Greene.

Another genre of illustrations hangs on the left wall: the already familiar diagrams of the Ark probably used as an architectural guide for building the Ark. Yet again, this detail presents Noah as a very skilled craftsman and engineer, able to engage in complex mathematical calculations and planning to construct an enormous vessel. However, combined with another artifact present in Noah's studio, these blueprints produce another important effect. Amidst these objects, which seem to be dislodged from their own time, the architectural drawings connect the assemblage of disparities to what is literally described in the Scriptures. They are the material reference to God's command that Noah should construct a three-floor ark (Gen. 6.16). The representation of an actual passage of the Bible is not displayed by chance. This setting has the function of materializing what creationists believe to be the real past. The reference intends to stabilize the chaotic narrative that the other artifacts, originating in different periods of time, provoke in the viewer's mind.

Not by chance, the studio is conceived like the ancient library of Alexandria, an inestimably valuable but also sadly lost historical treasure that might also have contained documents that described the beginning of the universe. The setting at the Creation Museum tells the audience that Noah was a literate man who had access to the original sources about the pre-Flood world. Like a biblical Herodotus, Noah is the privileged witness to great events and consequently the most trustworthy chronicler. Perhaps one day the actual documents he pens before viewers will be found, buried in a dusty library or museum or encased in receptacles and stashed in ancient caverns. Perhaps he is writing the missing pages of the book of Genesis. The material representation of Noah's wisdom and diligence not only assures viewers of his capability to construct the Ark and conduct its daily operation, but also lends authenticity to the Christian book. Sitting in his studio, Noah is represented as the author who *authorizes* the fidelity of the narrative of the book of Genesis.

Conclusion

At the conclusion of his article, Latour (1996) expounds why, in his opinion, the contemporary religious paintings exhibited in the Vatican Modern Sacred Art Museum are, in contrast to the Renaissance art displayed in the Pinacoteca Vaticana, of a 'nightmarish quality' (438). Apparently, new artists were never instructed about the principles and function of religious imagery. They fail to 'offer several layers of meanings [of interpretation]…' (437). 'If you paint a scene from the Bible, but without shaking its construction by inventing new indices that redirect attention away from it…then your painting will be more devoid of religious meanings than an oyster with lemon…' (438). What they simply propose is 'a vision of the practical production of facts' (436). Then, in his very last paragraph, he concludes: it is better to live in a civilization where scientific and religious images circulate, but each one in its own way. This clear separation of functions is definitely better than 'the rather horrendous culture in which the poor angels are harnessed to do the work of instruments, accessing the world beyond and carrying blank messages back to their return' (436). And here Latour almost seems to be referring to the scientific work performed by the diligent putti in Figure 2.

If we analyze the biblical settings in the Creation Museum and in the Ark Encounter according to Latour's perspective, the problem they posed is quite simple. They fail because they mix two distinct magisteria that should always be kept apart. They are of 'nightmarish quality' simply because they poorly convert biblical scenes into literal visual representations of a hypothetical past. In doing so, they do not accomplish the goal of providing 'several layers of meanings' and add important religious insights to the viewer. In a way what Latour is proposing is that if these sites look strange – in his own words, an 'hilarious attempt' – it is because they naively attempt to transform what is supposed to be a moral folktale into an historical and scientific narrative. Many would agree with that.

But such a simple verdict misses the fact that both sites are even more paradoxical than that. Far from being the products of an outmoded worldview, the scenes at the Creation Museum and the Ark Encounter represent a different class of religious imagery, a fascinating outcome forged in the inevitable encounter of the Bible with scientific and technological sensibilities. The conservative Protestant spaces indeed offer different meanings and religious insights because they show that instead of *sui generis* cultural phenomena, religions cannot be understood outside the milieu in which they exist. Fundamentalist Christians in the United States interpreted the Bible in concert with the imperatives of modern scientific sensibility. We should not dismiss this as an oddity but scrutinize the Creation Museum and the Ark Encounter as indices of just how deeply science has penetrated into religion. In order to make their scientific paradigm plausible, creationists must demonstrate that what is described in the book of Genesis really happened. Therefore, in the galleries of each site the Bible becomes a book that must be rationally completed. Creationists reject the idea of the events of the story of Noah propped up by an endless series of divine miracles. Accordingly, the Creation Museum and the Ark Encounter sustain the arch principle of the inerrancy of the Bible only by pushing its boundaries beyond the frail shape of an idea that is not native to the text itself. Carefully designed biblical scenes are not employed to teach moral lessons so much as to fill in the gaps and resolve the inconsistencies that riddle the ancient mythical tale. The irony cannot be lost on the reader: creationists are re-writing the Scriptures, using creativity and all manner of expensive devices to craft a version of the Sacred under the auspices of divine inerrancy.

Notes

1. To know more about Young-Earth Creationism or Creation Science and its relationship with the Bible as a book of factual records, see Whitcomb and Morris 1961.
2. For a discussion of these similarities, see Carneiro 2017.
3. I am indebted to an insightful presentation on the Ark Encounter by James Bielo (2015), and to his book *Ark Encounter: The Making of a Creationist Park* (2018), in which Bielo takes readers behind the scenes during the creative process of conceiving and building the Ark.
4. See, for instance, Dolkart 2009.
5. On Kircher's idea, see Godwin 1979. For discussion of Kircher's work on science and the Bible, see Breidbach and Ghiselin 2006.
6. For more information about Protestant visual culture, see Morgan (1999, 2007).
7. The Creation Museum also has a section that shows what happen to the world when people question the inerrancy of the Bible and replace Christianity with what creationists consider the menace of secularism. In this section, evolution, drug addiction, abortion and divorce, for example, are posed as results of a corrupted environment that has been systematically discrediting the veracity of the biblical timeline.
8. The book's second edition listed Constantin de Groot as the author. The first edition identified the author only as 'a lover of Jewish antiquities' (Liefhebber der joodsche oudheden). I am citing the second edition (de Groot 1700).
9. 'Ark planning', 2014.
10. 'Noah's Journey', 2014
11. 'Library of Ashurbanipal', n.d.
12. 'Noah's Studio', 2015.
13. 'Treasure from Desert Sands' (n.d.).

Bibliography

'Ark planning' (2014), Ark Encounter (blog), 23 May, https://arkencounter.com/blog/2014/05/23/ark-planning/ (accessed 16 August 2019).

Bielo, J. S. (2015), 'A Transmedial Bible: Performing Scripture, Producing a Creationist Theme Park', Presentation at the conference Material Religion: Embodiment, Materiality, Technology, Duke University, Durham, NC, 10–12 September.

Bielo, J. S. (2018), *Ark Encounter: The Making of a Creationist Theme Park*, New York: New York University Press.

Breidbach, O. and M. T. Ghiselin (2006), 'Athanasius Kircher (1602–1680) on Noah's Ark: Baroque "Intelligent Design" Theory', *Proceedings of California Academy of Sciences*, 57 (36): 991–1002.

Calmet, Augustines (1725–31), *Het algemeen groot historisch, geografisch, en letterlyk naam – en word-boek, van den gantschen H. Bijbel*. Translated by Mattheus Gargon. Leiden: Samuel Luchtmans.

Carneiro, L. (2017), 'Emulating Science: The Rhetorical Figures of Creationism', *Journal for Religion, Film and Media*, 3 (2): 53–64.

Dolkart, Judith F., ed. (2009), *James Tissot: The Life of Christ*, London: Merrel; New York: Brooklyn Museum.
Godwin, J. (1979), *Athanasius Kircher: A Renaissance Man and the Quest for the Lost Knowledge*, London: Thames & Hudson.
Gould, S. J. (1997), 'Nonoverlapping Magisteria', *Natural History*, 106: 7–21.
Groot, C. de. (1700), *Voor-bereidselen tot de Bybelsche wysheid, en gebruik der Heilige en kerklijke historien: uit de alder-oudste gedenkkenissen der Hebreen, Chaldeen, Babyloniers, Egiptenaars, Syriers, Grieken en Romeinen, tot eene merkelijke verligting der Goddelijke boeken*, Utrecht: Anthony Schouten & Hermannus Ribbius.
Kittler, F. (2010), *Optical Media*, Cambridge: Polity Press.
Latour, B. (1996), 'How to Be Iconophilic in Art, Science, and Religion?', in C. Jones and P. Galison (eds), *Picturing Science Producing Art*, 418–40, London: Routledge.
'Library of Ashurbanipal' (n.d.), *The British Museum Online Collection*, britishmuseum.org/research/collection_online/collection_object_details.aspx?objectId=303316&partId=1 (accessed 16 August 2019).
Morgan, D. (1999), *Protestants and Pictures: Religion, Visual Culture, and the Age of American Mass Production*, New York: Oxford University Press.
Morgan, D. (2007), *The Lure of Images: A History of Religion and Visual Media in America*, London: Routledge.
Morgan, D. (2012), *The Embodied Eye: Religious Visual Culture and the Social Life of Feeling*, Berkeley: University of California Press.
'Noah's Character Design – Noah's Got Skills' (2014), Ark Encounter (blog), 1 June, https://arkencounter.com/blog/2014/06/01/noahs-character-design-noahs-got-skills/ (accessed 30 September 2020).
'Noah's Journey' (2014), Answers in Genesis (website), 6 February, https://answersingenesis.org/bible-characters/noah/journal/ (accessed 16 August 2019).
'Noah's Studio' (2015), Ark Encounter (blog), 8 May, https://arkencounter.com/blog/2015/05/08/noahs-study/ (accessed 16 August 2019).
'Treasure from Desert Sands' (n.d.), *Antikensammlungen*, Munich, http://www.antike-am-koenigsplatz.mwn.de/en/ancient-masterpieces/museum-highlights/mummy-portrait.html.
Walker, S. (2000), *Ancient Faces: Mummy Portraits from Roman Egypt*, New York: Metropolitan Museum of Art.
Whitcomb, J. and H. M. Morris (1961), *The Genesis Flood: The Biblical Record and Its Scientific Implications*, Phillipsburg, NJ: P&R Publishing.

Chapter 10

Music, Scripture and the Sacred: Negotiating the Postsecular at a Dutch Arts Festival

Lieke Wijnia

Introduction[1]

Church buildings in Western Europe are increasingly the backdrop of cultural activities like exhibitions, concerts and festivals. This is not necessarily only an indicator of the demise of institutional forms of religion, but it can also be seen as a potential valuable transformation of religious sites and practices in secularizing societies. In addition to bringing external activities into the space of the church, a notable campaign in the Netherlands presents a collective of active religious sites as guardians of artistic and cultural heritage. Under the name *Dutch Museum Churches*, the campaign is initiated by the national museum for Christian art and heritage, Museum Catharijneconvent, which itself is housed in a former monastery.[2] Such campaigns are exemplary for continuous attempts at finding angles of relevance and urgency for religious sites in secularizing contexts.

Despite its purposes of preservation and passing on from one generation to the next, the heritage-angle faces the challenge of overlooking – or even ignoring – the living elements of religious traditions. Religious heritage does not only contain objects, practices and sites that were used in a far-away past, but these aspects just as much constitute practices in the present. The acknowledgement of this duality is one of the underlying principles in the emergence of the concept of the postsecular.[3] On the one hand, this theoretical notion reflects recognition of the drastic transformation of institutional forms of religion since the eighteenth century. On the other hand, it reinforces how the underlying motivations of which these institutions are the ultimate forms have not entirely disappeared, but instead are continuously transforming and additionally reappearing in alternative and new forms of equal significance and value. As such, in his review of the pluriform uses of the postsecular, Arie Molendijk described the relevance of the postsecular as the invitation to further develop the understanding of contemporary phenomena that constitute 'the "intertwinement" of the secular and the religious' (2015: 110). Whereas Molendijk argued that this intertwinement is based on one of the fundamental binaries with which humans understand their worlds (the religion–secular binary), I have argued that the relevance of the postsecular especially lies in the negotiations people enact to deal with (and partially overcome) this binary. The aim of the theoretical notion of the postsecular is not to identify respective religious and secular elements in particular cultural, social or political phenomena, but rather how within these phenomena people enact and embody religious and secular elements in mixed and merged forms – through ongoing negotiations (Wijnia 2018: 79–80).

In this chapter, I explore a case study that embodies this spirit of the postsecular, as it consists of a vast body of negotiations on religious and secular forms of the sacred. As such, this site of research is a prime example of Molendijk's intertwinement approach to the postsecular. Between 2012 and 2014, I conducted fieldwork at the annual Dutch arts festival *Musica Sacra Maastricht*. This festival programs a broad range of classical music, in church buildings and other types of monumental heritage sites in the southern city of Maastricht. Based around an annual theme, the concerts and performances are selected with the aim of exploring the sacred from a variety of perspectives. Looking through the lens of biblical tourism, it becomes apparent how this festival negotiates scriptural themes into a wider social relevance, while also acknowledging the historical and cultural significance of religious heritage in this originally Catholic part of the country. Even more so, through the selected music and other artistic performances, the festival connects two sides of the coin: of a transforming and living Christian tradition and of its rich cultural heritage.

To grasp the dynamics of the multi-faceted negotiations from which the festival results, this chapter particularly looks at the use of annual themes. Generally, these refer to the Bible, while the theme, which simultaneously functions as the festival title, can also be interpreted from a secular perspective or be connected to current affairs. Given that the theme guides the festival program committee's selection of music and other types of performance, understanding the multifold implications of the annual theme is highly relevant. To reach this understanding, first, the contemporary socio-cultural functions of the notion of festival are explored. This occurs particularly through the conceptual lens of ritual, emphasizing its symbolic significance. Approaching the festival as ritual demonstrates how, in the twenty-first century, festivals continue to draw from religious traditions, while simultaneously having gained a prominent place in popular culture. As such, the notion of festival itself can be seen as an intertwining form of religion and the secular. Second, the case study of *Musica Sacra Maastricht* is presented. Particularly, its history and transformation since its beginning in 1983 will be explored. This most impactful elements of transformation can be found in the name of the festival, the scope of its program and its manifestation in the broader cultural sector. All these elements contribute to the understanding of how the annual festival theme is employed in negotiations of the festival's postsecular identity. Third, via a brief exploration of recent theoretical approaches to the notion of the sacred, I position the selection of the annual festival themes within a broader framework. For the festival's program committee, this framework is mainly characterized by an inclusion of, but also expansion beyond, institutional religious traditions. How the program committee deals with Scripture in this balancing act between religion and the secular reflects the broader dynamic of negotiations of the postsecular. It is a balancing act resulting in the intertwined form that is *Musica Sacra Maastricht*.

The Festival as Ritual Practice

The festival is a ritual format that over time has transformed from being part of a cyclical, liturgical calendar to an omni-present cultural phenomenon. Festivals have become a fundamental part of the tourism industry. Traditionally they functioned as 'important forms of social and cultural participation, used to articulate and communicate shared values, ideologies and mythologies central to the world-view of relatively localized communities' (Bennett, Taylor and Woodward 2014: 1). The word *festival* is derived from the Medieval Latin *festivalis*, meaning 'of a church holiday'. In Old French it was used to describe something 'suitable for a feast, solemn, magnificent, joyful, happy' (Online Etymology Dictionary). In early times, the notions of feast and the festive were directly linked to liturgical calendars. With the rise of nation states in the nineteenth century, the festival was increasingly incorporated into the newly invented traditions and accorded civil rituals to give meaning and legitimacy to new forms of government (Hobsbawm and Ranger 1983). The notion and practice of the festival remained related to institutional religion, but also spread to other social and cultural domains. 'In a world where notions of culture are becoming increasingly fragmented, the contemporary festival has developed in response to processes of cultural pluralization, mobility and globalization, while also communicating something meaningful about identity, community, locality and belonging' (Bennett, Taylor and Woodward 2014: 1). As such, 'the contemporary festival therefore becomes a potential site for representing, encountering, incorporating and researching aspects of cultural difference' (1).

While the notion of festival generally brings to mind an image of rock, dance or pop music aficionados gathering in a temporary, especially designated festival terrain, classical music festivals are generally of a different order.[4] This type of festival consists of different modes of external expression of cultural identity, and alternative, more interiorized types of engagement with the performed music. Yet, the fundamental format in which music is perceived remains the same. Classical music has gone through what can be seen as a sacralization process throughout the twentieth century, which has brought particular ritual dimensions to its treatment in the concert hall. *Musica Sacra Maastricht* does not merely consist of a range of individual concerts, but actively tries to communicate its identity as a festival. While a festival may know physical expressions through its visitors as well as in concert locations, overall a festival has a rather virtual presence. It is not a *thing* that can be pinpointed, but a phenomenon that can be experienced and shared amongst its participants. The relation between festival and daily life, in terms of confirmation, inversion or abstinence, has been the subject of much theorization (e.g. Bakhtin 1984 [1965]; Gadamer 1986). Particularly, festival is characterized by being '"time out of time", (…) festival time imposes itself as an autonomous duration, not so much to be perceived and measured in days or hours, but to be divided internally by what happens within it from its beginning to its end, as in the "movements" of mythical narratives or musical scores' (Falassi 2011: 497).

The most tangible and material manifestation of the Maastricht festival, and of its identity, is the program booklet. This presents the parameters in which the festival takes place. Furthermore, the concept of *festival* takes shape through the notion of *space*. The space constituting the festival consists of two aspects; physical concert locations with their many cultural and historic meanings, and temporary affective spaces created by means of musical performances in specific architectural surroundings. The concert venues are carefully selected, primarily based on the acoustics and their location within the city. The types of buildings vary, but all have cultural-historic connotations. The concert locations may be generally divided in two types: those linked to institutional religion (churches, convents, chapels) and those with a different cultural-historic relationship, such as a factory hall turned into cultural space, the theatre and the city hall. The character and atmosphere of a concert space greatly contribute to the experience of the performed music. Music generates affective spaces not only through the performed sounds but also by means of how these sounds are related to the place in which they are performed. Affective spaces consist on the one hand of the formal qualities of the performed music and on the other hand of the impact this music may have on those listening to it, namely, the evocation of memories, emotions and associations. Affective space relates to both the internal and social worlds of the listener (Partridge 2013: 37-8). Music and the space in which it is performed simultaneously influence each other and imbue each other with meaning.

Although continuously transforming in relation to historical, social and cultural contexts, the festival practice retained its significance throughout history. Paul Post captured different elements of festive practices in a definition of feast:

> A feast is a moment or occasion on which people within the temporal order and at various stages in their lives, either individually or as a group or as a society, go beyond everyday life and in the form of a ritual give expression to events that mark the personal and social existence by means of a believing, religious or worldview orientation on meaning. (Post 1996: 35, and also Droogers 2001: 87)

This definition provides a lens through which festivals can be studied. Constituting this lens are four dimensions: a situation of contrast, performance of particular behavior, a dynamic between individual and collective identities and a reason why the festival takes place, a reason that brings all the participants together (Post 2001: 72-6). Within the context of the Maastricht festival, the contrast is created through the presentation of contemplative and complex music, in contrast to the fast-paced 'outside' world that is immersed in mass culture. This contrast requires particular behavior, most notably consisting of attentive and embodied listening during the performances. The identity dynamics taking place at the festival range between the individual listening to performances and the shared experiences of attending the festival with thousands of others. This results in a perceived shared valuation of particular forms of culture over and against other types of culture that are not included in the festival. And for the program committee the reason to bring people together is to contemplate the multifold manifestations of the sacred through music and other types of artistic performances. The next section discusses the history of the festival and the transformations of scope and aim since its inaugural edition, to explore how the various ritual dimensions feature in negotiations taking place in and around the festival.

From 'Religious Music' to 'Musica Sacra'

The festival now known as *Musica Sacra Maastricht* began in 1983 as *Festival for Religious Music*.[5] Its history demonstrates how the festival committee increasingly broadened the scope of the selected music. This, in turn, had strong implications for the resulting festival program as well as the locations in which the performances take place. With regard to its approach to the sacred, the festival history has resulted in a unique identity, both on national and international festival landscapes.

The festival began as part of a seven-yearly Catholic procession, the *Heiligdomsvaart*, which takes place in the city of Maastricht. The procession is held in the honor of the city's patron saint Servatius, whose relics are carried along the route throughout the historic city. The earliest record of the procession is from 1391. After this date, the procession was sometimes celebrated, but there were also times when the religious treasures remained locked away and no festivities were held. Since 1874 the procession has taken place every seven years, with the exception of 1944, when World War II prohibited it. Post-war, a new procession cycle began in 1948. Since then, the organization of the procession has been uninterrupted.

In 1983, the procession organizers felt that the single procession did not offer a sufficiently attractive program for the citizens and tourists in Maastricht and expressed a wish to expand its scope. So, the organizers of the procession asked the director of the Maastricht Cultural Centre if he was able to deliver a contribution to the procession activities in the form of a cultural, musical activity. The addition of religious music performances was considered a logical choice within this context, in line with the ritual character of the religious procession. It seemed a 'natural' extra activity (Interview with Theo Kersten 2012). In turn, this provided an extra dimension to the procession: in addition to the ritual practice, a touristic aspect was gained of attracting a broader and potentially returning audience. The initial *Festival for Religious Music* took place over two weeks, and each day had a luncheon and evening concert. After the first edition, it was decided to turn this festival into an annual event, taking place independently from the seven-yearly procession. After a couple of years, the focus shifted towards the weekends. The span of two weeks proved too long to attract sufficient audiences, despite many of the concerts being free of charge. In 1988, the name of the festival was changed to *Musica Sacra*. At the same time, the festival period was reduced to a weekend. The compactness and increased intensity were met by higher visitor numbers (Wijnia 2012).

For later editions of the *Heiligdomsvaart*, the festival committee would occasionally be asked for advice on musical contributions to the program of activities around the procession. The name change for the festival was initiated to allow for more programmatic space. It created the opportunity for the committee to select music dealing with religious connotations of the sacred, but also to select music that reflected other types of sacred concerns. The meaning of the term *sacra*, and how it is enacted in music, is continuously discussed among the program committee responsible for the festival program. In these discussions, which will be elaborated upon below, Rudolf Otto's *Das Heilige* (1917) is a frequent returning feature. That which invokes a combination of fascination and fear is sought after in the themes and music that is selected. The committee has set as its primary goal to offer possible explorations of the contemporary relevance of the sacred by means of the music in the festival program. The committee consists of employees of the theater that hosts the festival (*Theater aan het Vrijthof*) and its national media partner (broadcast associations *KRO* and *MAX* on the classical music station Radio 4).[6] In addition, a team of theater employees is responsible for the production of the festival program. During the period of my research, the committee members were all, save one, raised as Roman Catholics and biblically as well as philosophically interested and knowledgeable.

In addition to the broader scope of the musical section, over the years the festival has also become concerned with other artistic disciplines. In 2014, the official subtitle of the festival became *arts festival*, rather than *festival of religious music*. What first was an extra or parallel program has slowly but steadily become a fundamental and popular part of the festival weekend. These 'other' artistic activities consist of dance, film, theatre and visual arts. The local art house cinema offers a special program during the festival in line with the annual theme. Students of the Maastricht School of Theatre deliver theatrical contributions and students of first the Art Academy and later the Maastricht Academy for Design, Media and Technology, are responsible for visual arts contributions. For all students, the festival contributions are a part of their curriculum. All these program elements constitute an interdisciplinary approach to the notion of the sacred and the annual theme, as well as a widespread presence in the city during the festival period. The committee operates with the aim of living up to its interpretation of the festival name: a combination of *musica*, the old Greek term for the arts (not limited to music), and *sacra*, the notion referring to that which is experienced, valued and cherished as sacred (Meeting Program Committee, 9 January 2012).

Musica Sacra Maastricht has to continuously position itself in the Dutch festival landscape. The largest competitor in the field is the *Utrecht Early Music Festival* [*Festival Oude Muziek*], a ten-day festival of early music, covering the Middle Ages to the first half of the eighteenth century. In the public perception, and according to the festival program committee, the two festivals are frequently confused with one another, since the Maastricht festival also has its fair share of early music in the program. Moreover, the term *musica sacra* is often regarded as a synonym for early music. However, the program of *Musica Sacra Maastricht* also contains significant modern and contemporary segments. Because of this broad range of musical styles, in addition to its particular approach to the sacred, the festival has a unique identity within the Dutch context.

In the European context, other festivals have the term *musica sacra* in their name, but they are all different from the Maastricht festival. The bi-annual festival *Musica Sacra International* in Marktoberdorff, Germany, programs ritual and musical performances rooted in the world religions – Christianity, Judaism, Islam, Buddhism and Hinduism. It aims at interreligious dialogue and creating understanding for a range of cultures through music. *Festival Musica Sacra* in Austria, held in St. Pölten, Herzogenburg and Lilienfield, connects sacred music to the various services held on Sundays throughout the almost month-long festival period. Between 2003 and 2013, the weekend-long festival *Musica Sacra* took place in the Belgian town of Bever. In addition to a range of musical performances, the festival included a hike in the natural surroundings of the town. Traditionally, the festival closed with a performance of Arvo Pärt's *Kanon Pokajanen*. In the Swiss town of Fribourg, the *Festival International de Musiques Sacrées* is organized annually and consists of a week-long program of concerts of religious music from the Middle Ages to contemporary compositions. It has the same historical scope as *Musica Sacra Maastricht*, but stays focused on the religious, without exploring other varieties of the sacred. The focus on the contemporary relevance of the sacred as aimed for by the Maastricht festival can also be found in the *Festival de Fès des Musiques Sacrées du Monde* of the FES Foundation.

Its mission statement includes the aim of connecting different peoples and cultures by means of sacred music from all over the world.

A relatively recent initiative in this field is the *Lux Aeterna* festival in Hamburg, Germany, which saw its inaugural edition in 2012. Without having the term sacred in its title, this biannual festival in Hamburg appeals to a sense of transcendence by using the notion of eternal light. The subtitle *Music Festival for the Soul* implies an exploration of the idea that music contains the key to a heightened sense of fulfillment and meaning in life. The frame in which this festival is presented appears to be an attempt at adapting connotations of the sacred to a secular context. This festival seems to come nearest to the scope of *Musica Sacra Maastricht*, in that it tries to include but also reach beyond religious connotations in addressing the sacred.

Within the wider (inter)national festival landscape, for the program committee one of the defining features of the Maastricht festival's identity is the choice for the annual theme, its relation to the notion of the sacred and the manner in which the theme steers the selection of music and performances.

The Annual Festival Theme

The festival's change of name in 1988 provided a broadening of the possibilities in programming. To offer a red thread for both the program committee in their selection procedures, as well as for the festival audience in their selection of performances to attend, in 1990 it was decided to begin working with an annual theme (Table 10.1). The annual theme functions as a point of departure in programming artistic performances engaging with the sacred on some level. Despite the fact they are supposed to offer a red thread, the themes are often open for discussion and subject to multiple layers of interpretation. The themes always contain a relation to traditional religious vocabulary, while simultaneously linking to current topical social affairs.

Year	Translation in English	Original in Dutch
1990	Canticles	Hooglied
1991	Rites	Riten
1992	Requiem	Requiem
1993	Time of Suffering	Lijdenstijd
1994	The Greatest Beauty of Women	De Schoonste onder de Vrouwen
1995	Creation	Schepping
1996	Visionaries and Prophets	Zieners en Profeten

Year	Translation in English	Original in Dutch
1997	Saints and Idols	Heiligen en Idolen
1998	Apocalypse	Apocalyps
1999	Psalms	Psalmen
2000	Mysteries and Miracles	Mysteries en Mirakels
2001	Job	Job
2002	Of God and Gods	Van God en Goden
2003	Pilgrimage	Pelgrimage
2004	Holy War	Heilige Oorlog
2005	Angels and Demons	Engelen en Demonen
2006	Away from the World: Hermits and Monks	Weg van de Wereld – Kluizenaars en Kloosterlingen
2007	Visions of Eternity	Visioenen van Eeuwigheid
2008	Witnesses: Confessionals and Martyrs	Getuigen: Belijders en Martelaren
2009	Man and Woman He created Them	Man en Vrouw schiep Hij hen
2010	Devotion	Devotie
2011	The Joy of the Law – The Burden of Freedom	De Vreugde der Wet – De Last van de Vrijheid
2012	Rites and Rituals	Riten en Rituelen
2013	Introspection, Transformation, Conversion	Inkeer, Ommekeer, Bekering
2014	The Awe-inspiring	Ontzagwekkend
2015	The Way	De Weg
2016	Sacrifice of Love	Offer van Liefde
2017	In the Beginning	In het Begin
2018	Avenge, Forgive, Appease	Vergelden, Vergeven, Verzoenen
2019	To Pray, to Implore	Bidden & Smeken

Table 10.1. Annual festival themes between 1990 and 2019

Overall, in the festival's use of annual themes, three general types may be distinguished. Firstly, themes that directly refer to the Bible are used, such as Canticles (1990), Genesis (1995), Apocalypse (1998), Psalms (1999) and Job (2001). The themes of these Bible books are explored by means of the programmed musical performances. Their themes may have served as sources of inspiration for compositions, but can also be employed because they show parallels with performers or styles of performance. For instance, an Afghan lute player, who fled the Taliban, was presented as a contemporary Job. And the rappers of the Amsterdam group *Osdorp Posse* were framed as deliverers of contemporary psalmody, because they rap their visions of life on what appear to be simple melodic rhythms. In addition to direct Bible quotations, a second type of annual theme can be characterized as religious archetypes: Oracles and Prophets (1996), Saints and Idols (1997), Angels and Devils (2005) and Away from the World: Hermits and Monks (2006). This theme type offers the committee the opportunity to address societal and current affairs in the programming. For instance, in 2006 the focus was not only on historic hermits and monks but also on contemporary youngsters who feel more at home on the Internet than in public life. A third category consists of themes that refer to a certain type of behavior or practice, such as Rites (1991), Pilgrimage (2003) and Devotion (2010). At the beginning of the new millennium in 2000 an appropriate theme was chosen: Mystery and Miracles. The underlying thought was that despite technological inventions, mysteries would always be present. Every discovery implies new questions, a level of inexplicability, which the program committee regards as a parallel to what may be conveyed through experiences of music, notable but not always explicable.

Approaching the Sacred

The annual themes are used to offer a guiding hand in negotiating and exploring the domain of the sacred. Yet, the precise meaning of the sacred is subject to continuous debate, in both its practice and its academic research. This has resulted in a variety of approaches and the use of many different terminologies. A useful distinction is offered by David Chidester and Edward T. Linenthal, who distinguished what they called a *substantive* and a *situational* approach to the sacred. First, the substantive approach is employed mostly by theologians and religious studies scholars with a theological background:

> Familiar substantial definitions – Rudolph Otto's 'holy', Gerardus van der Leeuw's 'power', or Mircea Eliade's 'real' – might be regarded as attempts to replicate an insider's evocation of certain experiential qualities that can be associated with the sacred. From this perspective, the sacred has been identified as an uncanny, awesome, or powerful manifestation of reality, full of ultimate significance. (1995: 5)

Second, the situational approach to the sacred is employed by religious studies scholars, who aim to study religion, its practices and its beliefs, from a scientific distance:

> (…) [A] situational analysis, which can be traced back to the work of Emile Durkheim, has located the sacred at the nexus of human practices and social projects. Following Arnold van Gennep's insight into the 'pivoting of the sacred', situational approaches have recognized that nothing is inherently sacred. (…) [The sacred is] a sign of difference that can be assigned to virtually anything through the human labor of consecration. (Chidester and Linenthal 1995: 6)

Two recent approaches within the situational tradition are worth mentioning here, because they offer relevant theoretical parallels to the negotiations from which *Musica Sacra Maastricht* results. The first approach is offered by Gordon Lynch. He analyzed the sacred as communicative form (2012). His approach is positioned within the Durkheimian tradition, in which there is an intrinsic link between social and religious life. Lynch formulated a sociological approach to emphasize the conceptual relevance of the sacred in studying social dynamics and human behavior. This includes, but is not merely restricted to, the realm of religion. Lynch defined the sacred as 'a way of communicating about what people take to be absolute realities that exert a profound moral claim over their lives' (Lynch 2012: 11). By departing from the observation that forms of communication constitute the notion of the sacred, Lynch provided a direction for the analysis of the individual components of these communicative forms. 'They typically focus on specific symbols, invite people into powerful forms of emotional identification, and are made real through physical and institutional practices' (ibid.). The study of these components contributes to a larger understanding of the underlying dynamics of sacred forms. The second approach is formulated by Veikko Anttonen, who departed from an interest in language and the different usages of the term *sacred* in particular cultures over time. His research looked at 'meanings of the terms denoting "sacred" in different languages – present and past ones – and [studied] their relation to categories of cultural value in such contexts in which these terms have turned into religious concepts' (1996: 36). Furthermore, he argued that the notion of the sacred designates a universal dynamic in human behavior, however culture-specific the content of what or who is valued as sacred may be. Anttonen aimed to uncover a cultural logic underlying sacred-making behavior within a particular symbolic system. This resulted in the following description of the sacred:

> The sacred is a special quality in individual and collective systems of meaning. (…) Sacrality is employed as a category-boundary to set things with non-negotiable value apart from things whose value is based on continuous transactions. (…) People participate in sacred-making activities and processes of signification according to paradigms given by the belief systems to which they are committed, whether they be religious, national or ideological. (2000: 280–1)

Identifying the sacred as practices of categorization and valuation, Anttonen described a dynamic 'which at the same time "separates" and "binds"' (1996: 43). That which is regarded as sacred is separated because of the ultimate value attributed to it. Simultaneously, all those who uphold this valuation share it together, in turn resulting in a group dynamic, creating a sense of collectivity. The use of the term *sacred*, and the valuation this implies, suggests an immediate boundary between that which is sacred (and how it is supposed to be treated) and that which is not (which may possibly threaten the sacred). By designating the sacred as a category-boundary, Anttonen simultaneously directed attention to the realm of the profane, and to the sacred–profane interrelationship.

Lynch and Anttonen each approached the sacred in terms of its discursive character; the former in how communicative forms embody sacred values, which in turn impact morality; the latter in how the term is used and how this usage generates categories and value. Both approaches relate to negotiations from which the festival results, incorporating but also exploring beyond institutional religious discourse. The following question, then, becomes how such a broad approach to the sacred can be accessed and structured. If anything can be potentially sacred, how do we distill meaning from it? And how does this lead to a festival program? For the program committee, the tool to confront these challenges seems to be located in the use of an annual festival theme.

From Theory to Practice

In the realization of the festival program, several steps are followed. Two main questions are posed in the selection of concerts and other artistic performances: (1) *what is the link to the sacred?* And (2) *how does this performance (the music, performance style or composer's biography) fit the annual theme?* My field research covered the festival editions of 2012, 2013 and 2014. The treatment of the themes of these years is valuable in shedding light on the core elements in the committee's approach to the sacred. Although the presented discussions below are specific to this particular festival, they demonstrate the dynamics taking place in the field. As such, they have wider implications for understanding negotiations on the sacred which also take place in other socio-cultural contexts where the sacred is viewed through a lens of the arts.

First, the negotiations on the 2013 theme provide insight into how the program committee employs the notion of the sacred. Second, the discussion of the 2014 theme builds further on this approach to the sacred and how the dual approach of simultaneous inclusion and moving-beyond institutional religion is negotiated in practice. And, third, a discussion of the 2012 theme explores how negotiations taking place amongst committee members relate to those amongst the audience with regard to the appreciation of the sacred nature of ritual.

Introspection, Transformation, Conversion (2013)

The theme *Introspection, Transformation, Conversion* was approached in a pluralized manner. On the one hand, the theme referred to biographies of composers or persons that were sources of inspiration for these composers. On the other hand, it referred to the potentially transformative experience of listening to music. In addition to exploring different types of converts and composers who went through a transformation, the committee also aimed to offer musical performances that possibly lead to moments of reflection or deep thought. While wanting to refrain from saying what this reflection should look like, the committee's aim was to create conditions in which reflection take place as much as possible (Meeting program committee, 9 March 2012).

The challenge of this festival edition was how to demonstrate the notion of transformation in the concert programs. The only solution seemed to be to work with compositions that were written by a composer before and after his/her transformation and to see whether the musical style had changed. This was often discussed, but did not result in an eventual program segment.

Another relevant, and much-discussed, example was the work of Arvo Pärt, of whom the committee programmed his *Kanon Pokajanen* (1997) for the festival. At one point in his career, Pärt radically chose a strict way of composing. Yet, to note this choice and consequential change in his music required knowledge of his other compositions – knowledge that could not necessary be presupposed with all festival visitors. Simultaneously, the committee members regarded Pärt's musical compositions in terms of fostering reflection and transformation within the listeners. The pluralized approach to the annual festival theme took shape within one and the same composer and his body of work.

> Everyone who gets to know the work of Pärt, who listens to it for the first time, will be in shock. Can you listen to music like this, is this possible? And that makes it per definition, intrinsically, transforming music. And if you have listened to it a couple of times, it becomes familiar. Introspection is relative. (Meeting program committee, 14 March 2013)

More generally, this theme reflected how moments of silence and reflection were seen as prerequisites in constructions of the sacred. As stated in the 2013 project plan:

> In contemporary society, depth and reflection are strongly contrasted with everyday hastiness and economic capitalism that characterize us. The eternal struggle of man, to choose in enchantment and fear for matters that

transcend himself, resonates stronger than ever. Within this context, *Musica Sacra Maastricht* wants to contribute, by means of a compact thematic arts festival, to the spiritual life of mankind. (Projectplan 2013: 3)

This description contains an emic definition of the sacred (as operated by the festival's program committee): transcendent matters (events, people, experiences) that invoke a concurrent perception experience of enchantment and fear.

Awe-Inspiring (2014)

The 2014 festival theme was derived from scripture: *Terribilis est locus iste* (Gen. 28.17, 'Awesome is this place. This is nothing other than the house of God; this must be the gate to heaven'). This line reflects the simultaneous character of fascination with and fear of awe-inspiring matters. In its discussions during their meetings, the committee members often referenced the work of Otto, who defined the holy as a *mysterium tremendum et fascinans* (a terrifying yet fascinating mystery) (1958 [1917]). That which is too big to grasp is both fearful and attractive. The committee saw these elements as two crucial dimensions of the sacred. 'It is about mystery and inexplicability... Yes, but to not end up in the realm of mysticism we should maybe more focus on inexplicability. That which goes beyond us and what overwhelms us' (Meeting program committee, 24 May 2013).

The 2014 program brochure included an introductory statement written and agreed upon by the committee members. This introduction sheds light not only on the way the committee approached the awe-inspiring, but also on three fundamental aspects of the sacred as operationalized within the festival program. First, the contrast between the festival context and everyday life was emphasized. 'Everyday life usually offers little scope for surprise. But if we take the time to reflect on something, truly penetrating its depths, we come to some disconcerting discoveries and as many yet unanswered questions' (Musica Sacra Maastricht Magazine 2014). Second, this incomprehensibility of the cosmos, and the human place within it, was elaborated upon. 'The fact that we are increasingly better equipped to chart that cosmos, macro and micro, does little to diminish our realization that we are only a tiny part of an overwhelmingly large universe. We are amazed, perplexed and fascinated by what we observe, see or understand of the natural world around us and in us' (ibid.). And third, the limitations of individuals to gain control over their own destinies was reinforced. It concluded with an acknowledgement of a higher and stronger power than human beings can conceive.

And we are forced to admit that it is not us who are the dominating force; it is the powers that govern and shape us which are awe-inspiringly superior to us. We are in awe by the realization that although we may be able to divert and sometimes interact with these powers, we cannot control events like natural disasters, wars, disease, and death. This realization can provide the starting point for an ethical reflection about our actions. In any case, through the centuries it has produced some very meaningful art. That is exactly what *Musica Sacra Maastricht* is concerned with in 2014. (ibid.)

These quotations from the committee discussions and the festival brochure demonstrate a parallel between the committee's negotiations and recent sociological and cultural studies' theorizations of the sacred, as discussed above. This parallel rests in the broadened scope of the definition of the sacred. As a result, these explorations offer possibilities to open up traditional connotations (related to religious traditions and institutions) people may have of the sacred. Thus the festival hopes to appeal to a larger, both religious and secular, audience. The combination of definitional complexities and popular connotations results in continuous negotiations from the perspective of the program committee in its aim to serve the audience, while remaining true to its artistic and theoretical visions.

Ritual (2012)

The notion of ritual was a recurring theme during the committee meetings in the discussions about the selection of musical pieces. The committee often related the notion of ritual to the composition's structure or source of inspiration. The music's ritual character was only deemed to be realized when a composition was properly executed. Only then was rituality thought to be of a revelatory nature, distinguishing the performance from everyday routine. One of the music programmers described the annual theme as follows: 'Ritual not only marks important moments in our life, but also transcends it. The idea that ritual moves me beyond myself, relates me to something more than myself, is most important. The transcendental aspect is what we are looking for' (Meeting program committee, 5 April 2012). This experiential approach to ritual relates to the idea of the mystery of the sacred. The effects of ritual may be regarded as very evocative, and was described in terms that seem to draw from Romantic discourse. Yet, with regard to ritual musical structure (or a composer who employed a 'ritualistic mindset'), the committee members rather seemed to refer to music undone from all evocation and Romanticism. Another music programmer in the committee described this as the heritage of Stravinsky's *Sacre du Printemps*. To him this was music 'without personal feelings, anti-Romantic, strict, often in a very clear archaic structure. (...) It is a particular character, rather than that it has a function' (Interview with Sylvester Beelaert 2012). In addition to this experiential approach, the committee related the notion of ritual directly to the notion of liturgy. Music composed for a mass was per definition seen as music for a ritual context (Meeting program committee, 8 February 2012). Sometimes, the music was regarded as demanding a ritual context to become more substantial. In this argument, the committee discerned between performing music as a concert and performing it as a ritual (Meeting program committee, 23 April 2012). The intersection of concert and ritual performances produced a fine line between the notions of ritual and entertainment, between devotion and show.

An interesting case was the performance of a group of Shingon Buddhists. Their performances took place on three consecutive nights in the Protestant St. John's church. While most of the ritual practices took place at the church altar, along the church benches in the nave television screens were mounted, which announced the various parts of the ritual. These screens were seen as highly informative and useful for the audience. Simultaneously the addition of these devices reinforced how these particular performances in this particular context were not purely religious rituals, but rather a show in front of an audience unfamiliar with the ritual (and spiritual) connotations of the ritual gestures (Meeting program committee, 14 May 2012). In addition to the staging of the performances, the audience response was also contested. At the end of each performance on the first and second evenings, the audience did not respond with applause. The lack of applause indicated a perceived sign of respect for the ritual nature of the performance, and to reinforce that the audience did not regard this as an exotic show, but rather as a sincere ritual. Before the start of the third performance, the festival's project lead was told that the ensemble would certainly appreciate applause at the end this time, as token of gratitude; not necessarily for the artistic or theatrical merit of the performance, but for their presence and their sharing of the rituals with an audience (Meeting program committee, 10 September 2012).

This example demonstrates the contested result of a dynamic of ritual unfamiliarity, and uncertainty of the 'right' kind of response demanded by this ritual. In comparison, after the performance of *Sacrum Triduum Paschale* (rituals and chant from the Holy Week) by Schola Maastricht in 2012, people began to applaud after a little hesitation. The hesitant silence displayed respect for the ritual; the applause demonstrated appreciation for the singers. As this was a ritual context that was more familiar to the festival visitors, they were more certain in their convictions on how to behave respectfully – even though traditionally the proper code of conduct demands no applause in a church after the performance of a ritual (Wijnia 2013). The different responses to the performances demonstrate how different types of ritual evoke various negotiations. The negotiations between religious and secular notions of ritual (that of religious practices and secular concert attendance) resulted in either hesitant or no applause.

Concluding Thoughts

The annual festival theme is used as an instrument to create manageability in a broadened approach to the notion of the sacred and to find music that creates connections between scriptural or ritual significance and a sense of topical urgency. Biblical topics are related to current affairs, like the refugee crisis, social injustice or cultural intolerance. This results in a diverse festival program, including performances falling within the genre of *sacred music*, but also performances that are categorically positioned beyond it. The focus on the sacred turns the festival into a site that invites explorations of the relationship between artistic performances and potential interpretations of the sacred. Via the annual themes, the performances are placed within a framework of the pluriform face of sacred. Simultaneously, the committee members aim for the performances themselves to offer the festival visitors experiences that can potentially be characterized and valued as sacred. Within the festival context, the term *sacred* retains a problematic, contested and pluralized position, reflecting the contemporary position of religious sites and practices in a secularizing and globalizing world impacted by tourism as one of its shaping forces.

The festival committee needs the notion of the sacred in order to produce a culturally and socially relevant festival. It enables the committee members to select particular themes and select music that offers explorations of these themes. While the traditional connotations of the sacred are religious – predominantly Christian – the committee very deliberately wants to move beyond this frame. However, in the communication around the festival this provides quite the challenge. The only thing that may be of use is to continuously reclaim the theoretical potential of a broad approach to the sacred and how this translates into a variety of artistic practices. A Dutch music critic wrote about the 2015 festival edition: 'For a long time now, Festival Musica Sacra has presented much more, than only religious music, in a broad and diverse program' (Jansen 2015). More reviewers have mentioned the broad scope of the festival program over the years; yet, the use of the term *sacred* in the public domain continuously elicits initial connotations of a traditionally institutional interpretation of religion.

The recurrence of scriptural references in the annual themes constitutes a symbol of this direct link between religious traditions and practices and the sacred. The secular sacred is much less clearly defined, yet all that relates to institutionalized religion is automatically regarded as sacred. This, as I would call it, postsecular duality is operated with the intention of attracting a wider audience to the festival, including but also expanding beyond the religiously affiliated, knowledgeable and interested visitors. The broad approach to the sacred by means of the annual theme is also reflected in the selection of performance sites. By locating the performances in the city's many churches and heritage sites, two particular consequences are significant. First, a connection is established between these different types of sites and the variety of artistic performances located there. Second, new layers of significance for both the religious and non-religious heritage emerge during the performances. On the one hand, the festival continuously demonstrates how biblical themes – such as Canticles, Job or In the Beginning – have contemporary relevance, while on the other hand its program propagates the claim that the sacred can also be found outside of scriptural traditions. The festival's specific identity lies in the negotiations between these two dimensions, reinforcing its cultural relevance in a postsecular world.

Notes

1. This chapter further develops research I conducted for my 2016 PhD dissertation. The chapter sections build on the following passages in the dissertation: 'The Festival as Ritual Practice', 58–9; 'From "Religious Music" to "Musica Sacra"', 11–13; 'The Annual Festival Theme', 13–14/41–5/95–9; 'Concluding Thoughts', 174.
2. In Dutch the name of the campaign is *Het Grootste Museum van Nederland,* which translates as *The Largest Museum of The Netherlands.* It is interesting to note that the term 'churches' is not present in the Dutch campaign title, but only the notion of 'museum'.
3. Overviews of recent debates on the theoretical relevance of the postsecular are offered in Beckford (2012) and Molendijk (2015). Generally, philosopher and sociologist Jürgen Habermas is credited with popularizing the term. See Habermas 2001.
4. Although this is generally true, recent initiatives try to loosen up the identity of classical music festivals. An example is the *Wonderfeel* festival, in format modeled after popular festivals, taking place in a natural park in the north of the Netherlands. Visitors can set up their tents and camp there for the duration

of the festival, an activity generally linked to festivals like *Lowlands, Pinkpop* or *Rock-Werchter.*
5. The festival has been organized under the following names. Festival of Religious Music (1983–1984); Euro Festival of Religious Music (1985–1987); Musica Sacra (1988–2001); Musica Sacra Maastricht (2002–present).
6. At the time of writing this chapter, due to organizational shifts in the division of airtime on classical radio station Radio 4, it is unclear how the festival will partner with national media in future editions.

Bibliography

Anttonen, V. (1996), 'Rethinking the Sacred: The Notions of "Human Body" and "Territory" in Conceptualizing Religion', in T. A. Idinopulos and E. A. Yonan (eds), *The Sacred and Its Scholars: Comparative Methodologies for the Study of Primary Religious Data*, 36–64, Leiden: Brill.

Anttonen, V. (1999), 'Does the Sacred Make a Difference? Category Formation in Comparative Religion', in T. Ahlback (ed.), *Approaching Religion*, 9–23, Turku: Scripta Instituti Donneriani Aboensis.

Anttonen, V. (2000), 'Sacred', in Willi Braun and Russell T. McCutcheon (eds), *Guide to the Study of Religion*, 271–82. London and New York: Cassell.

Bakhtin, M. (1984 [1965]), *Rabelais and his World*, Bloomington: Indiana University Press.

Beckford, J. A. (2012), 'SSSR Presidential Address: Public Religions and the Postsecular: Critical Reflections', *Journal for the Scientific Study of Religion*, 51 (1): 1–19.

Bennett, A., J. Taylor and I. Woodward, eds (2014), *The Festivalization of Culture*, Farnham: Ashgate.

Chidester, D. and E. T. Linenthal, eds (1995). *American Sacred Space*, Bloomington; Indiana University Press.

Droogers, A. (2001), 'Feasts: A View from Cultural Anthropology', in P. G. J. Post, G. A. M. Rouwhorst, L. van Tongeren and A. Scheer (eds), *Christian Feast and Festival: The Dynamics of Western Liturgy and Culture*, 79–96, Leuven: Peeters.

Falassi, A. (2011), 'Festival', in C. T. McCormick and K. Kennedy White (eds), *Folklore: An Encyclopedia of Beliefs, Customs, Tales, Music, and Art*, 493–501, Santa Barbara: ABC Clio.

Gadamer, H. G. (1986), *The Relevance of the Beautiful*, London: Cambridge University Press.

Habermas, J. (2001), *Faith and Knowledge*, Acceptance Speech Peace Prize of the German Book Trade.

Hobsbawm, E. and T. Ranger, eds (1983), *The Invention of Tradition*, Cambridge: Cambridge University Press.

Jansen, K. (2015), 'Met Zachte Trom op Weg naar het Einde', *NRC Handelsblad*, C7, September 22.

Lynch, G. (2012), *On the Sacred*, Durham: Acumen.

Molendijk, A. L. (2015), 'In Pursuit of the Postsecular', *International Journal of Philosophy and Theology*, 76 (2): 100–115.

Musica Sacra Maastricht Magazine (2014), *Ontzagwekkend*.

Online Etymology Dictionary, Entry 'festival', https://www.etymonline.com/search?q=sacred (accessed 28 August 2013).
Otto, R. (1958 [1917]), *The Idea of the Holy*, Oxford: Oxford University Press.
Partridge, C. (2013), *The Lyre of Orpheus. Popular Music, the Sacred, and the Profane*, Oxford: Oxford University Press.
Post, P. (1996), 'Liturgische Beweging en Feestcultuur. Een Landelijk Onderzoeksprogramma', *Jaarboek voor Liturgie-Onderzoek*, 12: 21–55.
Post, P. (2001), 'Introduction and Application: Feast as a Key Concept in a Liturgical Studies Research Design', in P. G. J. Post, G. A. M. Rouwhorst, L. van Tongeren and A. Scheer (eds), *Christian Feast and Festival: The Dynamics of Western Liturgy and Culture*, 47–77, Leuven: Peeters.
Projectplan Musica Sacra Maastricht 2013, Maastricht: Theater aan het Vrijthof.
Wijnia, L. (2012), 'Festival Musica Sacra Maastricht: Van het Eerste Uur', in *Jubilee Magazine Musica Sacra Maastricht – Rites and Rituals*, 18–22, Maastricht: Theater aan het Vrijthof.
Wijnia, L. (2013), 'Religious Rituals as Festival Performances at Musica Sacra Maastricht', *Yearbook for Liturgical and Ritual Studies* 29: 99–111.
Wijnia, L. (2016), 'Making Sense through Music: Perceptions of the Sacred at Festival Musica Sacra Maastricht', PhD diss., Tilburg University.
Wijnia, L. (2018), *Beyond the Return of Religion: Art and the Postsecular*, Leiden: Brill.

Chapter 11

BUILDING ON THE GOSPEL:
THE MORAVIAN SETTLEMENT AT CHRISTIANSFELD

Marie Vejrup Nielsen

This chapter explores the question of the Bible and tourism through a case where the Bible is present at a cultural heritage site in the life and practices of a religious community, the Moravian Church in Christiansfeld, Denmark. Christiansfeld was founded as a Moravian settlement in 1773 in the midst of controversy surrounding pietistic Christianity in Northern Europe. In 2015, the town was recognized as a UNESCO World Heritage Site (United Nations Education, Scientific and Cultural Organization). This study explores the intersection between cultural heritage management and religious identity and practice by focusing on the scriptural structures on which the town is built. The chapter adds further perspectives to the study of the Bible and tourism by emphasizing the presence of biblical heritage in the form of material culture and ritual practices, which reflect biblical ideals. This perspective is developed here through combining perspectives from the study of lived religion with those from the study of religion and tourism. This is then used as a framework to analyse the nomination document from the UNESCO process and the ritual of the Love Feast (a bi-annual event celebrating the renewed Moravian Church), which are cases of the intersection of cultural heritage and religion.

**Theoretical Perspectives:
Lived Religion as Cultural Heritage**

Religion, tourism and cultural heritage is a research field with many different perspectives (Stausberg 2011; Timothy and Olsen 2006).

Heritage studies is an ever-growing field, as attested, for example, by *Palgraves Handbook of Contemporary Heritage Studies* (Waterton and Watson 2015). Another entry point for the study of religion and cultural heritage has been the study of religion and museums. Here scholars have provided important insights into the question of religious objects and museums (Paine 2000, 2013; Buggeln, Paine and Plate 2017).

This chapter will focus on the intersection of religion, cultural heritage and tourism, and will therefore not refer in any detail to the vast amount of literature on cultural heritage generally. It will connect contemporary perspectives from the study of 'everyday religion' and 'lived religion' with perspectives from studies of tourism and religion.

Researchers within the study of contemporary religion have called for a shift away from a previous focus on the official dogma of institutional religion. Instead, they urge researchers to examine religion in the non-official, lived, everyday life of ordinary people (McGuire 1992; Hall 1997; Ammerman 2007). This turn towards the lived religion of everyday life also meant a turn away from official texts as interpreted by the elite towards those materials that expressed the religious perspectives of the non-elite. Instead of examining holy texts, such as the Bible, in their dogmatic, institutional context, now objects, foodstuffs, music and movements throughout the spheres of life became the relevant data for researchers. In recent years, there has been a call for a study of lived religion within institutions (Ammerman 2016; Nielsen and Johansen 2019). This chapter will discuss how Scripture can be expressed through materials such as buildings and town plans, as well through the religious practices of a community, that is, as part of the lived religion of a community.

This study draws on perspectives already applied in research connected to UNESCO World Heritage sites, since it combines the study of the key documents in the appointment process with fieldwork. This approach is at the core of the extensive work on UNESCO World Heritage by Michael A. Di Giovine (2009). In his work, Di Giovine develops the concept of Heritage-scape to describe the complex dynamics of developing and maintaining the global heritage management system of UNESCO. Di Giovine underscores how '[…] a place is inscribed as a World Heritage site not because it is something, but rather because it is representative of something that can be understood, in part, through touristic interactions with the place' (Di Giovine 2009: 39). Drawing on Clifford Geertz, Giovine studies the ideological claims of the heritage scape, as they are 'Textually expressed in narrative form and materially manifested through monumentality […]' (119). This chapter follows the approach of Giovine in its focus on expressions of the foundational narrative of the specific religious community of the Moravians in Christiansfeld as it is represented in the official documents of the UNESCO process, and the interaction between tourists and congregation at the ritual of the Love Feast.

Cases of living religious communities and cultural heritage have been studied in various contexts (Geary 2017; Shepherd 2013). In *Faith in Heritage* Shepherd discusses the issue of UNESCO World Heritage and religion in relation to the specific case of sacred Buddhist site Wutai Shan in China. He focuses on how the Western ideals of heritage management play out in a Chinese context. Shepherd's work underscores the importance of studying what happens when what is a religious site for some becomes cultural heritage for all. He points to how religion included in cultural heritage sites is 'simultaneously preserved and vibrantly alive' (Shepherd 2013: 13). When a religious community is part of a site that receives the status of World Heritage, it enters the scene of global tourism. Shepherd points to how, in the case of Wutai Shan, there is a high degree of complexity in the various positions among the agents involved at the site. The case that Shepherd analyses is in itself very different from the case analyzed in this chapter, but he formulates key research questions for the study of living religion within cultural heritage sites: 'Whose heritage is being preserved and managed by whom, for what purposes, and with what political implications?' (24). In this chapter, it is not the political implications that are central, but the implications for the lived religion of the community and how their identity is represented in the context of cultural heritage.

Michael Stausberg provides an extensive overview of many aspects of the study of tourism and religion (Stausberg 2011). His book, *Religion and Tourism*, presents a range of entry points for the study of lived religion and tourism at religious sites that have become tourist attractions. Stausberg explores the concepts of 'tourist' and 'pilgrim', and challenges the juxtaposition of the two, as visitors to a site do not necessarily conform uniformly to the one of the other (64). Stausberg points to how '[…] by their sheer physical presence, international tourists pose a challenge at the sacred places. Exchanging gazes and even taking part in rituals may threaten the purity of the site or endanger the religious economy. At the same time the tourists are a powerful new market force […]' (69). There are many interests at stake, and neither the tourists nor the religious practitioners of the site can be said to represent uniform interests. In addition, Stausberg concludes that, 'The interface between tourism and pilgrimage results in processes such as accommodation, dispute, negotiation and resistance' (71).

A central element in Stausberg's perspective is a focus on the spatial dimensions of tourism (75) and how in the case of tourism to religious sites, this means the construction of a shared space. In addition, in examining tourist motivations for visiting these sites, he points to how:

> [...] in tourist settings, accordingly many religious places are generally not primarily visited for their specific religious qualities, i.e. as places to perform rituals or create religious experiences, but more often because of their apparent or assumed history or historical significance and their aesthetical dimensions, including the architecture, arts, their sheer size, and other spatial features [...]. (77)

Stausberg discusses the UNESCO World Heritage list specifically, and how this 'global canon of the worlds most prestigious heritage sites' aims at securing the protection and continued management of sites that are deemed as having value for all of humankind: 'World heritage sites belong to all the peoples of the world, irrespective of the territory on which they are located' (97–8). And if a living religious community is a part of such a site, it might become what Stausberg describes as a 'religious communities on display':

> However, in many cases it is not so much the doctrinal content of their religious identity, but their particular, clearly distinguished and, judged from mainstream society, seemingly anachronistic lifestyles grounded in their religious doctrines and identity that warrant their status as tourism attractions. (152)

Stausberg then discusses cases where this has led to friction between religious communities and tourists, and consequently more direct attempts to instruct tourists on what to wear, how to behave and where to go and not go. Stausberg points to how the reaction of tourists who happen to be around 'when something happens' (170) can have an influence on the experience of the congregation, as they are often 'staying just for a short while, with attention primarily on the aesthetics of the space' (172).

Stausberg's perspectives underscore how the intersection between religion and tourism is a complex field with many different possible experiences and consequences. Other scholars have examined cases of the intersection between religion, tourism and space. In *Tourism, Religion and Spiritual Journeys*, both Thomas S. Bremer and Daniel Olsen address the importance of space and place (Timothy and Olsen 2006). Bremer points to how, when a religious site is visited by tourists, 'a simultaneity of places emerges in parallel geographies of both the sacred and the touristic.

Moreover, the meaningful content of these overlapping places relies upon narratives of identity that consolidate the bonds that religious adherents and tourist visitors feel toward the place' (Bremer 2006: 25). Olsen addresses the specific challenges of management of a site: '[…] the main focus of religious site custodians is to preserve the emotive qualities of the place as a way of creating and maintaining an atmosphere conducive to worship and contemplation' (106). The religious identity of the space, constructed and maintained, in some cases by the continued religious activities at the site, is at the core of the value of such sites.

These theoretical perspectives all aim at examining the complex intersections between religion, tourism and cultural heritage, and the questions they raise form the basis for the primary questions of this chapter:

- How does the intersection between religious identity, cultural heritage and tourism express itself in relation to the site of Christiansfeld, which is a case of a religious community on display?
- How is the simultaneity of being preserved and vibrantly alive, as formulated by Shepherd, apparent in this case?
- In addition, what is the dynamic between the doctrinal content of their religious identity and lifestyles, grounded in their religious doctrines and identities, expressed in relation to their value as cultural heritage and tourism attraction for all on a global scale?

These questions will be answered with a specific focus on Christiansfeld as an example of how the biblical text is present at a cultural heritage site in the lived religion of a religious community, and by examining the material manifestations of the scriptural structures on which the town is built.

Data and Methods

The material analyzed in this chapter consists of elements of a larger research project concerning the lived religion of the Moravian Church in the context of a cultural heritage site. Here we will analyze textual material, primarily the document from the UNESCO World Heritage Nomination process, but supplemented with material from the current UNESCO website and a regional tourist management website. The analysis of the religious ritual of the Love Feast is based on participant observation at four such events held at Christiansfeld from the reopening of the Church Hall in 2016 (November 2016; August 2017; November 2017; August 2018). The participant observation consisted of participation in the preparation of the Love Feast together with the servers, here primarily the female servers. I took part in the work together with the female servers of preparing the event, but during the ritual, I joined the other participants on the pews. During this work, I engaged in conversations with the servers, and was able to observe practices and conversations as they took place at the site, including interactions between tourists and servers. During the participant observation, I was primarily with the female servers and to some degree the male servers, which means that I did not observe the preparations made by the choir.

Christiansfeld:
Historical Background and Contemporary Situation

Christiansfeld was founded as a Moravian settlement in 1773 in the midst of controversy surrounding pietistic Christianity in Northern Europe.[1] The settlement was part of a global religious movement, where the establishing of specific townscapes was a key expression of faith. Situated for a period in the Central European region of Moravia, or Mähren, as it is named in German, which today is part of the Czech Republic, the Moravians arrived in Denmark at a time of strict control of religion. At the same time, new thoughts and ideas were spreading throughout Europe, including within religion, such as the pietistic call for a more personally committed Christianity (Bach-Nielsen 2012: 283). The advisor to the king at the time sought to reform Danish society, and saw the Moravian settlements as key to this, due to their industrious work ethic (Mettele 2005). Pietistic impulses had reached Denmark, inspired by developments in Halle and the arrival of pietists in Copenhagen, including the controversial Count Nikolaus von Zinzendorf. Count Zinzendorf had allowed a group of refugees with a pietistic faith to settle on his lands in Saxony, which in the 1720s became the first Moravian town of Hernnhut. On the Sunday closest to the 13th August every year the Moravian Church celebrates the founding event of the Moravian community in 1727. Zinzendorf was a celebrity in his own time, both famous and infamous for his involvement in radical Christian milieus (Fogleman 2007). The idea of establishing a colony, a Hernhutt on Danish soil, went forward, despite the controversies. The king finally offered fell land belonging to the royal estate Tyrstrupgård. The rights of the new community were stated in a royal concession, which also gave it ten years of tax exemption. The town was constructed so as to be self-sufficient, and the Moravians who arrived at the settlement were artisans, carpenters, masons, brewers and bakers. Together they built houses and constructed the necessary frameworks for a Moravian life. The religious ideas behind establishing towns can be traced back to the early years of the community, and the town plan of Hernnhut became the master plan for all Moravian settlements. The primary idea centered on the ideal religious community, sustained through community with the crucified Christ, who was their leader in all things. Each town should clearly express this, and have Jesus Christ as the ordering principle. There were therefore no mayors or other forms of civil or 'secular' leadership in the town; it was governed by the councils of Elders. In Christiansfeld, as in the other Moravian colonies, these ideas manifested in buildings and a town plan, with every move guided by these ideals.

The center of the town is the Church square, which is surrounded by the central buildings: the Church Hall and the Choir Houses. The Choir Houses were originally housing for unwed men and women, one house for each group. The gender division found in the town plan is a key example of how the materiality of the town is an expression of religious ideals. The Sister House is placed to the north, as is the Widow House, and the Brother House is situated in the south. This is a reflection of the Church Hall, where sisters sat in the North side and brothers at the South side of the Hall, and entered the church by two separate entrances. The terminology of sisters and brothers is central to the Moravian faith and is the central identity marker throughout life.

The life of a member of the congregation was a life of movement around the town according to the religious ideals. Young, unwed men or women lived in the designated housing. Family life was lived in the family housing, and if one became a widow or widower, gender-divided housing was provided. Even in death, the Moravians maintain the primary identity of sisters and brothers. This is visible at the cemetery at Christiansfeld called 'God's Acre', referencing the biblical imagery of death as a stage upon the road to the final resurrection, the harvest. The funerary culture of the Moravians is that sisters and brothers are buried separately in designated areas of the cemetery in the order of death, not, for example, husband and wife together as was and is generally the norm in the burial culture of Denmark. The stages of life are laid out in the town plan and the same goes for the everyday life, through the schools for children and the small factories, shops and workshops, such as the carpenters, bakery and brewery. The Council of Elders oversaw all industry. The result of the first phase of the establishment of Christiansfeld was visible in the new fully functional town, where every aspect of life was intended to express the religious ideals: a town built on the Gospel.

Today, the majority of the original houses are still owned by the congregation, but the traditional pattern of life as a movement between buildings in the townscape is not fully maintained; for example, gender division within the society is not upheld, meaning that the housing according to life stages is not maintained. However, in the religious and ritual context key elements remain as part of the lived religion. The congregation consists of approximately 150 members in the town and 200 living elsewhere in Denmark. In 2015, Christiansfeld was included on the list of UNESCO World Heritage sites, after having been included on the Danish tentative list since 1993. The congregation, through the council of Elders and the pastor, were very active in the process of nominating Christiansfeld.

We will now move to an analysis of the key examples of how the intersection between cultural heritage and living religious identity of a community is represented in the nomination document and on the official UNESCO website, before looking in more detail at the celebration of the Love Feast.

Christiansfeld: Nomination and Status as World Cultural Heritage Site

The complex processes behind the nomination for inclusion on the list of World Heritage Sites will not be discussed in detail here. Instead, the focus will be on the final stage of the process, the nomination itself, as well the representation of Christiansfeld as World Heritage site by UNESCO. The nomination document, *Danish World Heritage Nomination – Christiansfeld – a Moravian Settlement* is an official document and the result of a long process (Nomination 2014). It was prepared by an official steering committee consisting of Sven Felding (Danish ICOMOS (International Council on Monuments and Sites), Bolette Lehn Petersen (The Danish Agency for Culture), Käte Thomsen (Representative of the Moravian Church in Christiansfeld, member of the Elder Council), Jørgen Bøytler (Pastor, Moravian Church in Christiansfeld), Hans-Jørgen Bøgesø (Advisor) and four representatives from Kolding Municipality: Thomas Boe, Lone Leth Larsen, Karen Stoklund and Annemette Løkke Berg. The steering committee therefore had members both from the official, political system on the regional level, as well as members representing the congregation on a leadership level, both lay and clergy. The document consists of 455 pages, and the final page carries the official signature from the state of Denmark, in this case, the residing Minister of Culture.

The primary goal of the document is to convince the UNESCO evaluators that this site should be included on the World Heritage List. It is therefore a document that reflects the UNESCO criteria.[2]

The nomination document begins with a presentation of the designated area of the site: the original town plan area, including the cemetery (*Danish World Heritage Nomination, Christiansfeld – a Moravian Settlement*, Kolding Municipality, 2014: 13). The maps and descriptions are followed by the core argumentation of how Christiansfeld meets the criteria of World Heritage. Two general criteria were used in the nomination: 'Criteria (iii): a unique or at least exceptional testimony to a cultural tradition or to a civilization which is living or which has disappeared'. And 'Criteria (iv): an outstanding example of a type of building, architectural or technological ensemble or landscape which illustrates a significant stage in human history' (14). The steering committee did not use the criteria that specifically mentions living traditions and belief (criteria vi).

The two selected criteria function as the benchmarks for the rest of the nomination document. This is done through a link between the religious identity and its materialization in the townscape:

> Christiansfeld is an example of a Protestant ideal city. The town today presents an intact and well-preserved structure and collection of buildings. The town plan consists of two East–West oriented tangential streets around a central square and a cemetery placed outside of town. The town also reflects the Moravian Church's societal structure, which is characterized by large communal houses for the congregation's widows and unmarried men and women. […]. The architecture is homogenous and unornamented, with one- and two-storey buildings in yellow brick and with red tile roofs. The proportions, materials and craftsmanship contribute to the town's special atmosphere of peace and harmony. (14)

And:

> All of the details of Christiansfeld were tailored to ensure that members of the Moravian Church could lead lives in accordance with the denomination's Christian and cultural values. Christiansfeld thus also presents an exceptional connection between town structure and denominational culture. (15)

The nomination also connects the religious identity and life of the congregation today to the value of cultural heritage. This is linked to the themes of authenticity, protection and management:

> One special strength is that the town's architecture exists as a greater whole, reflecting the Moravian Church's fundamental philosophy and desire to live in a fellowship of moderate and peaceful Christian life. This philosophy is evident in the detailed craftsmanship and the materials selected for the buildings. (15)

And:

> The buildings, however, are still in use, mostly as residences but also for small business and shops located in the nominated area. Congregational life in Christiansfeld is still very active. The members of the congregation maintain their religion within the town and its buildings, honouring old traditions and making new ones. The interaction between the well-preserved collection of buildings and the continuance of Moravian Church life in the town grants Christansfeld exceptional authenticity. (16)

The connection between the religious identity of the Moravian Church and its material manifestations is marked throughout the nomination. The scriptural structures on which the town is built are held up as having value as cultural heritage for all. This is also visible in the presentation of the cemetery: 'God's Acre (God's Acre is the field where the Lord shall one day reap)' (45). And,

> The inscription on the main gate leading into God's Acre is also noteworthy: 'It is sown in corruption'. This is a scriptural quote from the text read during a funeral procession leading the dead to rest. On the inner side of the gateway are inscribed the words: 'It is raised in incorruption', so that the mourners may depart God's Acre with the message of the Resurrection ringing in their ears. (46).

Here, the scriptural reference is not explained through its textual context; instead, it is situated as the scriptural foundation for the lived religious practices of the community. Scripture is manifested in the practices of the community in the form of the material and immaterial cultural heritage of the cemetery.

The strong connection between lived religious life and material expressions is presented throughout all sections of the nomination. The buildings and their materials are 'All reflecting the Moravian Church's frugal philosophy' (50). And: 'The same forms and varieties of stucco, door paneling, and fittings are present in many of Christiansfeld's buildings, bearing witness to a thematic strictness in accordance with the Moravian Church's spiritual ideals' (50).

In the section on the 'Moravian Church today', the changes throughout the long history of Christiansfeld are addressed, and contemporary life is described through the key value of community. Here tasks performed by the congregation are at the center:

> There are other tasks as well, including rubbish collection, clearing of the attics, pruning of trees, assistance at God's Acre, lawn mowing, packing and distribution of the congregational flyer in Christiansfeld and its surroundings, custodial duty of the museum, recording of the museum objects, and cleaning of the museum. Other things that contribute to the congregation's sense of community are the Congregation Councils, lectures, Bible Hour, study groups, sewing groups (without or without actual sewing) [...]. (76)

The maintenance of cultural heritage is brought forward as a key element in relation to the museum. The dual values of preservation and vibrantly lived religious life, as presented by Shepherd, are here bound together into an argument for the status of the site: the religiously formed ideas as maintained by the lived religion of the community are represented as a perfect fit for the management ideals of preservation.

The scriptural structure of the town plan and buildings is represented as reflected in the attitude of the community in their work ethic: 'The Moravian Church's culture is expressed through the congregation's work with preserving its buildings, where the denomination's faith and philosophy of simplicity is reflected in the town's architecture, craftsmanship, and the social relationships [...] (78). This is further explained through the concept of 'liturgical life':

> [...] this perspective implies that, in a Christian's life, there can be no differentiation between secular life and the church. Existence cannot be divided into secular and spiritual zones. Human life consists of work life, private life and church life, with Jesus serving as role model. The activities of daily life, – even sleep, – are liturgical actions. All actions are guided by the Saviour and shaped in accordance to His will. (123)

> Jesus' life is regarded as a model for the life of every Christian, and all life thus becomes a single unit in which all of humanity is interwoven: the religious and the secular become impossible to entangle because all life belongs to God and must be spent in his service. (123)

A final element of the nomination will be mentioned here before moving on to the Love Feast. The images that fill the pages of the nomination are, in the majority of cases, professional photos of the town, with some historical photos. Throughout the nomination, the materialities are in focus: the buildings, the beautiful rooms with their austere, clean white and wooden surfaces, impressive in their simplicity. People are rarely portrayed in the images, and only on very few occasions are there identifiable contemporary Moravians, the exception being one image alongside the description of the Love Feast, where a female server is seen (80).

Throughout the document, the congregation is represented as uniquely equipped to take care of this site, not only because they are the custodians of history, but also because they are infused with a preservation logic stemming from their understanding of religious community and service as expressed through service in all aspects of life. Stausberg mentioned how tourist settings, including many religious places, are 'generally not primarily visited for their specific religious qualities [...] but more often because of their apparent or assumed history or historical significance and their aesthetical dimensions, including the architecture [...] (Stausberg 2011: 77). The nomination document circumvents this dichotomy, by arguing that the aesthetic, architectural dimensions are expressions of religion, and that the religious dimension is part and parcel of taking in the beauty of the austere, simple aesthetic of the buildings, rooms and townscape. The ideal of not separating secular from religious spheres in the lived religion of the congregation is translated into cultural heritage management language by linking this religious identity directly to the town as a material manifestation of religious identity. The 'emotive qualities of the place', as presented by Olsen, are activated to a very high degree as a quality created by the religious ideas and practices of the community (Olsen 2006: 106).

The argumentations of the nomination proved convincing, and in 2015 Christiansfeld was placed on the World Heritage list. The official UNESCO today states that:

> The Moravian Church community remains very active in upholding its religious and social services. These also form opportunities for involvement in the social and ethical principles that underline the significance of the settlement. The structure and characteristics of the original town plan remain largely unaltered. All buildings, especially those of the early Moravian period of 1820, retain their authenticity in material, design, substance, workmanship, and some of them as well in function and use. The continuity of the Moravian Church community contributes to safeguarding authenticity in spirit and feeling as well as atmosphere of the property. (UNESCO website)[3]

This evaluation clearly states the points of value from the perspective of the global, heritage management system: the congregation ensures the value of the site by ensuring authenticity through spirit. It is cultural heritage for all, on a global scale. And its value is strongly connected to the religious ideals of the congregation as they are manifested through material culture, primarily in the form of the town plan and its buildings. The evaluation is focused on the materiality of buildings and their historical value, but also recognizes the link between the congregation and the town plan and buildings.

The Love Feast: Celebrating Community

The Love Feast is the celebration of the founding events in 1727, the beginning of the renewed Moravian Church. It is a musical service, where the choir and participants sing selected verses that address the relationship between Christ and his community. During the songs, the sisters and brothers who works as servers, dressed in traditional outfits and using traditional materials, serve sweetened tea and raisin buns. This is done as a remembrance of how the brothers and sisters in 1727 had been so caught up in the experience of community through song and prayer that they did not eat, which led Count Zinzendorf to send tea and raisin buns to sustain them. The servers serve the pastor first, and then move along the pews. Male servers serve the men's side of the Church Hall and female servers the women's side. When performing the ritual role, the servers become very visible as carriers of the historical practices and ideals, both for congregation members and for tourists or other visitors.

The Love Feast is traditionally held twice a year, marking the founding of the church on 13 August 1727, and the day of the Elders on 13 November, celebrating the date when Jesus Christ was elected General Eldest of the Moravian community in 1741. Through the Love Feast, the congregation expresses the core elements of its religious identity: the community with each other and with Jesus Christ. The Love Feast is an example of a religious festival that is also a shared space, but not necessarily a shared place for tourists and members of the congregation.

This chapter will focus on the day of the event and on the expressions of religious identity through a special focus on interaction between tourists and servers. The analysis of the interaction is a valuable vantage point, as it provides important insights into the actual interaction at the shared site at ground level, beyond the statements of official documents and political agendas. It provides a window into how members of the congregation and visiting tourists behave in this common space in relation to a religious practice, which is deeply rooted in the biblical structures and religious identity of the Moravian congregation.

The atmosphere of the last preparations is one of busy, focused work carried out together in a positive frame of mind. The dress code is central to the ritual. The sisters and brothers have historically worn their Sunday clothes for this service, and today this is the uniform of the servers. That is, today, what was once the general dress code has become only ritual dress. The men wear black suits and black shoes, and those serving wear white gloves. The women wear a black dress, covered with a lace apron and servers wear gloves. The women wear a special lace cap, which has ribbons in different colours. Originally, the colours signified stages of life: unmarried virgin, married and widow. They bring their own black dresses, but the aprons, caps and gloves are in most cases kept by the congregation in rooms at the Church Hall.

When observing the dressing of the women, the expressions of identity and the transformation of life during the centuries become visible. Everyone put on aprons, caps and gloves together. When dressing, there is a lively debate as to how to wear the aprons correctly, and memories of mothers and grandmothers are also consulted. This often brings up stories of the servers' own memories of participating in the events. The ribbons offer one context for the discussion of the transformation of the lived, everyday life of the congregation and the traditional religious ideals, in that the colors of the ribbons no longer match the modern way of life of the female servers. This causes light-humored conversations, for example in relation to whether a divorced woman is more an 'unmarried virgin' or 'a widow', since no historical category is available. This is one example of the negotiation of tradition that takes place behind the scenes in the congregation.

The preparation of the tea and the setting up of the cups is a central element of the planning of the Love Feast. The water for the tea is boiled in the original kitchen and the tea is served in special cups, handled with care, and when we were placing the cups on the trays, the history of the cups was brought up. The cups are, like the dress code, an item that is transformed into a material expression of the identity of the congregation in the ritual. Next to the rows of cups, room is made for the porcelain tea pots and for the big copper tea pots. The buns are baked based on original recipes and are kept in large wicker baskets from where they are served during the Love Feast.

During the preparations, the doors are open to allow for the many activities not only of the servers, but also of the choir. This means that visitors will sometimes enter the room, in a case of what Stausberg described as 'passing by when something happens' (Stausberg 2011: 170). Tourists enter, walk around, take photos and also approach those preparing the event. The same happens after the Love Feast, where often the servers are busy again, since there is often a preparation for a new service the same day.

Here one feature becomes clear for the observer. Once the servers are dressed, they become visible as 'Moravians' in a unique way. When visiting Christiansfeld as a tourist, there is no way of telling whether a person you meet on the street is a Moravian. At others sites, where a specific cultural heritage has died out completely and is only maintained as re-enactment, tourists might encounter museum staff dressed up to reenact life as it once was. Alternatively traditional outfits might have been maintained in everyday life, as with the Amish. In Christiansfeld there is neither re-enactment nor maintenance of traditional outfits in everyday life. So, when entering the Church Hall during the preparations for the Love Feast or during and after the service, tourists encounter the rare sight of an 'identifiable Moravian', who is clearly engaged in something authentic and interesting. Naturally, the tourists are drawn to the servers.

Some of the tourists will address the servers, and ask question such as 'what are you doing?' and 'what is going on in here?'. The servers are very busy at this point, but they do take the time to answer the questions. The servers express to me that everyone is of course welcome as participants in the Love Feast. The main concern has to do with the ideals of participation over against tourists, who might just pop in, take a picture and leave. There was a strong preference for the tourists or visitors to take part throughout the Love Feast; to sit down and join the community. Thereby the servers express the ideal of a participating community, which is at the core of the religious life. In relation to the perspective brought up by Stausberg, that outsiders participating in rituals 'may threaten the purity of the site or endanger the religious economy' (Stausberg 2011), this case points to how the opposite can also be the case; e.g. the non-participation in a ritual, which is about community, can challenge the ideals of the site. In this case, the strong values of community, building on the Gospel, can be challenged by passive bystanders.

The servers did not see themselves as tourist guides or information staff; they are carrying out a central role in the life of the community. At the same time, they uphold ideals of hospitality and openness. The Visitor Center had not taken on the role of guiding visitors during 2016–2017 in relation to the specific event – they primarily offered the information online. During 2018, a new model was tried out, where the pastor gave a brief introduction to the Love Feast for anyone interested, prior to the event. This initiative was developed as a collaboration between the congregation and the Visitor Center. The talk was held in an adjacent building, the Sister House, which also houses the museum and the Visitor Center. One of the sisters, who was also serving the Love Feast the same day, participated and showed the special dress worn by the female servers. This development is of course small in itself, but it does point to how the interaction between cultural heritage management and congregations develops in light of the intersection between tourism and religious life of the congregation.

Discussion and Conclusion

The analysis of these examples of the intersections of religious identity, cultural heritage and tourism in the case of Christiansfeld provides the background for a discussion of the representation of a specific religious community as cultural heritage for all.

The lived religion of the congregation is front and center in the nomination document in relation to the connection between religious ideas and practices on the one hand, and the criteria for being cultural heritage for all of humanity on the other. Christiansfeld was represented as a town built on the Gospel, and maintained as such through the ideals and practices of the congregation. The scriptural structures underpinning the lived religion of the congregation were also present in the preparations of the Love Feast. Here, there was an emphasis on community and the continuation of the life of the congregation, expressed through official ritual and the behind-the-scenes preparations.

Stausberg describes the variety of intersections between tourism and religion as 'accommodation, dispute, negotiation and resistance' (Stausberg 2011: 71). The nomination document could be seen as a case of accommodation between a religious identity and secular criteria of a global, heritage management institution. However, it is not an accommodation in the form of a resignation of primary elements of the religious identity to the demands of an external institution. Instead, it is the reformulation of core values and practices in the context of cultural heritage. Stausberg pointed out that tourists will primarily visit Christiansfeld for the aesthetic and historic experience, and here it is interesting to note that the nomination document embraces this by translating the central religious identity into a language that embeds the aesthetics in a religious context. The town is beautiful and relevant as cultural heritage because of the continued Moravian way of life, not as a fossilized, historical re-enactment.

The nomination document argues for a complete overlap between the ideals of the religious community and the ideals of cultural heritage management, thereby arguing for an overlap of places in a form that resembles the discussions by Bremer (2006: 25). This chapter does not look at tourists' perspectives in themselves, but it shows how the same questions of distinct spaces and shared places can appear in the representation of a living religious community in the context of cultural heritage management.

At the same time, the strong emphasis on community as participatory and active become the key point of potential tension between the congregation and tourists. The same ideals, which are held up as part of global cultural heritage and ensure the continued preservation of the values of the site, come into play when busy servers, preparing a central event, are called upon by tourists to act out the role of guide and source of information. or when the carefully prepared event is interrupted by tourists entering and leaving without engaging with the ideals of participatory community.

The congregation does not see itself as re-enacting a religious ritual for the sake of spectators. It is part of their own living and lived religious tradition, expressed through their tasks of preparation for the ritual. At the same time, the new status means that there can be a higher influx of visitors from outside the local community at some events. The congregation, who have also been a key part in the process of being recognized, does not necessarily see this as a problem. However, it leads to new challenges. The Love Feast has elements that are often seen as marketable in other tourist contexts: it is an exotic ritual, which entails special dress, special food and drink, movements and music. The question is whether these cases of intersection between tourists and congregation will lead to larger changes in years to come; some changes have already begun, for example the new idea of having a short introductory lecture for tourists prior to the ritual.

The official heritage management institution, UNESCO, has recognized the values as they were argued in the nomination document, as well as recognizing that the site is special because of the congregation's continued, lived religious life, expressing not only the historical dimensions of Christiansfeld, but ensuring its continued value for all the world as cultural heritage. The religious life of the congregation, including its 'liturgical work' where the maintenance of buildings and townscape is at the core of congregation life, was represented as constantly infusing Christiansfeld with value. In the nomination document, this translates as a description of the way the religious identity of community with

Christ, here and in the hereafter, is expressed throughout the lives of the Moravians. The core understanding of the town as built on the Gospel, that which is religious heritage and lived religion for some, is translated into the language of cultural heritage and thereby represented as valuable global heritage for all.

Notes

1. This chapter uses the official English name of The Moravian Church. The Danish name, Brødremenigheden, directly translates as 'Congregation of brothers' or 'The Brethren'. The church identifies itself as having its roots in Christian reform groups prior to the Lutheran reformation that existed for a time in Moravia (Mähren).
2. Source: whc.unesco.org/en/criteria/ (accessed 16 August 2019).
3. Source: whc.unesco.org/en/list/1468 (accessed 16 August 2019).

Bibliography

Ammerman, N. ed. (2007), *Everyday Religion: Observing Modern Religious Lives*, Oxford: Oxford University Press.

Ammerman, N. (2016), 'Lived Religion as an Emerging Field: An Assessment of its Contours and Frontiers', *Nordic Journal for Sociology of Religion*, 29 (2): 83–97.

Bremer, T. S. (2006), 'Sacred Spaces and Tourist Places', in T. J. Dallen and D. H. Olsen (eds), *Tourism, Religion and Spiritual Journeys*, 25–35, Routledge Contemporary Geographies of Leisure, Tourism and Mobility, New York: Routledge.

Buggeln, G., C. Paine and S. B. Plate, eds (2017), *Religion in Museums: Global and Multidisciplinary Perspectives*, London: Bloomsbury Academic.

Carsten Bach-Nielsen, 'Pietisme mod og med staten', in vol. 2 of Carsten Bach-Nielsen and Jens Holger Schjørring (eds.), *Kirkens Historie*, 283–316, Copenhagen: Hans Reitzels Forlag.

Di Giovine, M. (2009), *The Heritage-scape – UNESCO, World Heritage and Tourism*, Lanham: Lexington Books.

Fogleman, A. (2007), *Jesus Is Female: Moravians and the Challenge of Radical Religion in Early America*, Philadelphia: University of Pennsylvania Press.

Geary, D. (2017), *The Rebirth of Bodh Gaya – Buddhism and the Making of a World Heritage Site*, Seattle: University of Washington Press.

Hall, D. D., ed. (1997), *Lived Religion in America – Toward a History of Practice*, Princeton: Princeton University Press.

McGuire, M. (1992), *Religion: The Social Context*, Belmont, CA: Wadsworth Thomson Learning.

Mettele, G. (2005), 'Kommerz und fromme Demut. Wirtschaftsethik und Wirtschaftspraxis im "Gefühlspietismus"', *VSWG: Vierteljahrschrift für Sozial- und Wirtschaftsgeschichte*, 92 (3): 301–21.

Nielsen, M.V. and K. H. Johansen (2019), 'Transforming Churches: The Lived Religion of Religious Organizations in a Contemporary Context', *Journal of Contemporary Religion*, 34 (3): 509–27.

Nomination (2014), *Danish World Heritage Nomination, Christiansfeld – a Moravian Settlement*, Kolding Municipality.

Paine, C. (2000), *Godly Things: Museums, Objects and Religion*, Leicester: Leicester University Press.

Paine, C. (2013), *Religious Objects in Museums: Private Lives and Public Duties*, London: Bloomsbury.

Shepherd, R. J. (2013), *Faith in Heritage: Displacement, Development, and Religious Tourism in Contemporary China*, Walnut Creek: Left Coast Press.

Stausberg, M. (2011), *Religion and Tourism – Crossroads, Destinations and Encounters*, New York: Routledge.

Timothy, D. J. and D. H. Olsen, eds (2006), *Tourism, Religion and Spiritual Journeys*, Routledge Contemporary Geographies of Leisure, Tourism and Mobility, New York: Routledge.

Waterton, E. and S. Watson, eds (2015), *The Palgrave Handbook of Contemporary Heritage Research*, New York: Palgrave.

Afterword

James S. Bielo

In September 2019, the *New York Times* reported that Saudi Arabia would begin issuing three-month tourist visas in 2020 for the first time in the Kingdom's history. One month later, a related story circulated through U.S. conservative media networks. The Christian Broadcasting Network (CBN), for example, told website visitors that this new opportunity in global travel included 'access to ancient biblical sites, including "the real" Mount Sinai'.[1]

For centuries, Jebel Musa in northeastern Egypt has been the purported location of the scriptural mount Sinai revered in Jewish, Christian and Islamic traditions. The Egyptian site is an established pilgrimage destination, home to a sixth-century Orthodox monastery, an eleventh-century mosque and a UNESCO World Heritage site since 2002. Recently, fundamentalist Protestant writers have challenged this history.[2] Alternatively, they argue that Jebel al-Lawz in northwest Saudi Arabia is the true site where Moses received the 10 Commandments.

The CBN story promoted a U.S.-based tour company, Living Passages, who planned to offer the 'first-ever Christian tour[s] of rare sites' beginning in February 2020. Living Passages has led group tours since 1996 – 'Christian travel for the intrepid Berean', their tagline proclaims.[3] Interested travelers can choose from Israel (note: not Palestine), Jordan, Egypt, Ethiopia, tracing 'the footsteps of Paul' on a Mediterranean cruise, sites from the life of Martin Luther in Germany, a creationist safari in South Africa and now, Saudi Arabia. The tour leader, a conservative Christian pundit and self-styled 'security analyst', tweeted a video of himself at Jebel al-Lawz to his 42,000+ followers in December 2019. The short promotional video promised potential travelers a 'safe' and 'affordable' journey to the 'real' Mt. Sinai, 'sanctioned' by the state and led by 'Saudi guides'.

Living Passages' organized entry to Saudi Arabia represents something of a perfect storm of late modernity and Protestant fundamentalism. A nation-state structurally invests in global tourism as an economic strategy. A tour company with established international ties promotes in the register of 'pioneer', more so than 'pilgrim' or 'tourist', reviving Orientalist fantasies of biblical discovery that fueled the rapid expansion of travel to Palestine in the nineteenth century (Rogers 2011; Kaell 2014; Kugelmass, this volume). And, a claim to authenticity is made in the dual registers of scientific legitimacy ('biblical archaeology') and conspiratorial suspicion of non-Protestant tradition. As Living Passages explains in its description of the tour: 'though [Empress Helena] never set foot in the land, she claimed that mount Sinai was in the Egyptian peninsula. No one dared to question the queen for hundreds of years, and as a result, this became the traditional site of mount Sinai on the Sinai Peninsula.'[4]

Destination Bible

The case of Saudi Arabia's mount Sinai joins a long history of travel that this volume calls biblical tourism: travel oriented around scriptural texts, stories, tropes and interpretations. If you are a contemporary traveler interested in being a biblical tourist, you have a lot of options. You might seek out sites mentioned in sacred Scripture, taking you to modern-day Israel-Palestine, Egypt, Jordan, Turkey, Lebanon, Syria, Iraq, Iran, Greece and Italy. As the example of Living Passages and Wilensky-Lanford's chapter illustrate, depending on your theology or your interpretation of evidence, other nations join this list. But whose Scriptures are we talking about? If you are an African Hebrew Israelite, biblical tourism might take you further south into the African continent (Jackson 2013). If you are a Mormon, as Olsen and Pierce illustrate in their chapter, biblically framed travel might include stops throughout the Americas. And, as Dein's chapter illustrates, Chabad Jews who scripturalize the writings of the late Rebbe Menachem Schneerson make pilgrimage to his burial site in Queens, New York City (cf. Wimbush 2015).

Searching for other options? There are many. As the volume makes clear, biblical tourism is more expansive than just the sites named in Scripture. At least since Helena's fourth-century romp through Palestine, items tied to biblical history have been transported worldwide (Wharton 2006). You might seek out the *Scala Santa* in Rome, the 28 steps Catholic tradition says are those that Jesus ascended to meet Pontius Pilate; the *Santa Casa* in Loreto, where Catholic tradition says the House of the Annunciation resides; the Glastonbury thorn (Barush 2018); or, quite literally, thousands of other proclaimed relics of human bodies, sacred shrines and natural landscapes displayed throughout the world (cf. van Asperen, this volume; Paine, this volume).

Still room in your itinerary? If you seek out sites that have been commemorated as historically significant – heritage-ized, as it were – then your list of possible stops expands significantly. Flush with time and money? You might consider Shaker Village of Pleasant Hill, a preserved nineteenth-century community in central Kentucky; or Lalibela, a series of twelfth-century Ethiopian Orthodox rock-cut churches in the Amhara region; or festivals celebrating the Protestant Reformation in Wittenberg, Germany (Stephenson 2010); or Christiansfeld, an eighteenth-century Moravian settlement in central Denmark (Nielsen, this volume).

Keener for Marian sites? While the Vatican has authorized only 16 apparition sites – in Rwanda, Poland, Lithuania, Mexico, Ireland, Italy, the Czech Republic, France, Belgium, Portugal and the U.S.A. – hundreds of others have been authorized by local bishops, Coptic tradition, or thrive through popular authorization despite official disavowals (e.g., Badone 2007). Your travels would take you to more than 70 nations, throughout the Americas, Europe, Africa and Asia.

Just as interested in replications and imagined re-creations? This might blow your travel budget. Marian apparition sites like the healing grotto of Lourdes in France have been re-created hundreds of times over. Biblical stories – from Noah's ark to Solomon's Temple, the Wilderness Tabernacle and scenes from the life of Jesus – also have hundreds of re-creations in miniature and 'life-sized' form (Bielo 2017).

While some destinations become non-extant, others undergo revitalization and still others are created or 'discovered', such as Saudi Arabia's mount Sinai or a first-century materialization of Nazareth in a D.C. museum (Paine, this volume; Bielo 2020a). If this all seems never ending, that's because it is. Taken together, the eleven chapters in this volume offer a revealing glimpse into this wildly diverse phenomenon, and provide instructive conceptual frameworks and research methodologies for apprehending its historical and ongoing significance. In closing, I offer a further conceptual reflection, one oriented toward the global mobility of people, materialities, stories, ideologies and charisms.

Circulation

Of the thousands of sites we might include under the rubric of biblical tourism, each and every one of them has some kind of local resonance. They are situated in particular places, engaged by people in those local settings and embedded in specific cultural, political, economic and bureaucratic contexts. The chapters collected here do a masterful job of illuminating local resonances, such as Wijnia's analysis of the postsecular environment shaping a Dutch arts festival and Nielsen's exploration of the ways in which residents and travelers experience Christiansfeld. Here, I pivot away from themes of local histories and encounters and call our attention toward the social life of biblical tourism across places (cf. van der Beek, this volume). What if we conceptualized the mass of the world's biblical tourism sites as an uncoordinated infrastructure of and for circulation? What if we conceive of individual sites as material nodes in an ever-productive network of movement?

'Circulation' is a load-bearing term in this proposal, and I am drawing inspiration from Handman (2018). Working from a linguistic anthropology standpoint, her concern is with 'the ways in which Christian speech and interpretive frameworks move through time and space' (153). While her focus is on practices and processes of Christian evangelism, the framework she develops is applicable more broadly. It captures a pivotal dynamic in any religious culture that engages public spheres, drawing attention to the trails of circulation, the actors involved in making materialities move and the institutions and conditions that structure movement. By understanding circulation, we better understand how social worlds are 'infused with the communicative norms and materials forms' of religious traditions (158; cf. Engelke 2013). Along with a dizzying array of other media – audio and audio-visual technologies from CDs to podcasts, print and digital texts from tracts to postcards – places can also figure as 'media of circulation' (153).

Similar to almost any site of tourism-pilgrimage, biblical tourism sites are productive: generating new travelers, literature and other promotional resources, purchased and collected souvenirs and stories of experience. These products move beyond the physical confines of the site, going out into the world while always looking back to their origin as an invitation to 'go there'. Places and their circulating elements are embedded within broader processes of authorization, in which the promised power of a site is cultivated and performed through an assemblage of media forms, actors, relations and technologies (Morgan 2014). In turn, places are part of a networked infrastructure that circulates the enchantment of sites, promising visitors something special, from access to divine presence to miraculous healing, enriched religious sociality and ideological claims to territory and/or legitimacy.

What, then, circulates from and among biblical tourism sites? One answer, perhaps most obviously, is people. Choreographed itineraries can move individuals and groups from one site to another, as with tours that follow the missionary journeys of St. Paul as an authorized European 'cultural route' (Moira, this volume). As pilgrim-tourists move among sites, they bring with them templates for being an onsite visitor. It may also be worth asking how choreographies for experiencing place also circulate among sites. Nielsen's analysis of Christiansfeld (this volume) poses a useful example, in which visitors come interested in being participant consumers in Christian heritage. To what extent is this experiential model co-produced in an emergent way onsite, and to what extent is it imported from other sites of history-making?

Another answer is objects in various forms, also not too surprising given material culture's importance for pilgrimage (e.g., Coleman 2014). As several contributors here observe, multiple apparati of travel (e.g., guidebooks, pendants) circulate with and beyond travelers (Kugelmass, this volume; van Asperen, this volume). Purchased souvenirs are widely regarded as a revealing kind of materiality whose social life can move from commodity to gift to keepsake (Kaell 2012). A less widely observed example is landscape items; elements removed from nature to be circulated as a gift or memento (van Asperen, this volume; Olsen and Pierce, this volume; cf. Kaell 2014; Bartal, Bodner and Kuhnel 2017; Bielo 2018). In the case of landscape items derived from scriptural places, they circulate because of the potency attributed to them as physical pieces of biblical history. At New York's Palestine Park, a daughter of the Chautauqua Institution's president in the 1920s returned from a pilgrimage to Palestine with a vial of Dead Sea water, which she placed in the park's replica equivalent (Long 2003: 37). Nearly a century later, in 2014, the Temple of Solomon opened in Sao Paulo, Brazil: a neo-Pentecostal complex featuring an outsized replica of the Hebrew Bible temple, accompanied by Wilderness Tabernacle and Garden of Gethsemane replicas. It boasts eight million dollars' worth of Jerusalem stone as part of the Temple's exterior construction, four date-palm trees from Israel-Palestine and a dried bush from Egypt's mount Sinai.

A less common form of object circulation is the movement of organized displays from one site of biblical tourism to another. In the U.S. context, dozens of homespun and entrepreneurial projects have started only to go non-extant for various reasons: from energetic founders passing away to legal disputes. In some cases, the displays deteriorate in place or become refuse, but in many cases, they are moved to new homes. The example of 'Bible Walk' presents one such scenario.[5]

In 1977, a charismatic Catholic named Bill Warren received a 'very strong prompting' to create a series of biblical statues in a forest. He and his wife Gail followed the prompting, purchasing around a hundred fiberglass figures and 40 acres in a town southwest of Pittsburgh, Pennsylvania. They arranged the figures, clothed in 'biblical dress', into 30 scenes on a quarter-mile pathway in the woods. Four years later, the site was closed because of a legal challenge from the local government that the area was not properly zoned for tourism. Bill remembers the dispute skeptically, in both secular and spiritual registers. He said that the owners of a golf club planned for nearby did not want tour busses and other tourist activity, and so bribed local officials to shoo Bible Walk away. 'Satan', Bill explained, 'was fighting hard to disrupt anything that could bring people to the Lord'. The fiberglass figures dispersed; Warren donated half to a Friary in West Virginia and half to a church in Mansfield, Ohio. The Friary closed a few year later and the Ohio pastor retrieved the remaining half, opening Bible Walk in 1987 in a new building adjacent to the Mansfield church (see Eyl 2019).

If we imagine places on the landscape as relatively enduring and mass-produced objects as relatively ephemeral, the movement of organized displays beyond the life of their original location flips this temporality. Things can outlast places, circulating long after a place has been removed from a landscape, perhaps only alive through documented or narrated memory or perhaps erased altogether.

Stories, one way to narrate memory, are yet another example of circulation. Volume contributors illustrate well the power of stories in drawing visitors to sites of biblical tourism, as well as engaging other affective processes: remembering, memorializing, sacralizing and publicizing. The chapters by van der Beek and Dein are especially rich in this regard. In her analysis of the Camino's global circulation, van der Beek illuminates how stories cohere across media sources and perform the socio-historical work of secularizing. To the extent that the Way of St. James has changed from being a strictly biblical pilgrimage to a heritage-ized resource and journey for secular and spiritually ecumenical travelers, the stories that van der Beek engages with are integral to this change. In his analysis of the *Ohel*, the Lubavitcher Rebbe burial site, Dein examines the circulation of miraculous healing stories and their cumulative effect of enchanting the place while also authorizing the scripturalization of the Rebbe's writings.

The process of sites generating stories that travel widely is a circulatory thread stretching across time. Van Asperen's medieval travelers to Loreto, Walsingham and Wavre carried and traded stories as well as pewter badges. And the nineteenth-century guidebook writers considered by Kugelmass were nothing if not purveyors of stories. In 2017–18, I conducted ethnographic and archival work with a site called the Garden of Hope in northern Kentucky (Bielo 2019). Opened in 1958, this site has experienced numerous custodians and cycled through periods of revival and neglect. For Steve, the primary tour guide since 2003, the Garden is a wellspring of stories, many of which he shares with visiting tour groups. One of Steve's favorites is about a mysterious stranger who rescued the Garden's founder by solving a vexing constructing problem. Physically imposing and oddly dressed – 'six foot six, 360 pounds, bib overalls, a straw hat and no shoes' – the stranger appeared out of nowhere and disappeared without a trace. The story of the man (angel?) in bib overalls is part of a typed script for guides that Steve inherited when he started leading tours, though it did not originate there. It was circulating in print for at least a decade, first appearing in a popular devotional book in 1992, and referenced in newspaper and periodical profiles of the Garden beginning in 1995. As much as authorized guidebooks or other media of cultural production, circulating stories play a pivotal role in establishing and publicizing the identity of scripturalized places.

The final answer I will put forward to the question of what circulates from and among biblical tourism sites is a meta-semiotic observation. Sites circulate as signs, used to index broader phenomena and concerns. In this way, sites are polysemic resources, used for diverse purposes by divergent actors. They are mobilized as evidence of inspiring devotion and of religion-gone-wrong. The Creation Museum in northern Kentucky, represented so deftly in this volume by Carneiro's chapter, circulates as the public face of creationism, as the *sine qua non* for understanding this fundamentalist movement's identity and ambitions. The circulation of sites as signs can also reveal competing frames used to render places meaningful.

Consider the example of Holy Land, U.S.A. in Waterbury, Connecticut. This site was started in 1956 by a then-60-year-old Catholic lawyer named John Greco. Situated on a hilltop above a small industrial city, the site grew to a collection of over two hundred miniature replicas of biblical stories. Greco's Holy Land was a mélange of discarded materials. No single archival source lists every material used, but the following inventory emerged from reading across sources: plaster, concrete, statues of saints, pews and stained glass from churches, lumber, tin, plastic sheeting, plastic plants, marble, chicken wire, stone, brick, scrap metal, aluminum, copper, cement, radio and television cases, hand rails, soup pots, stovepipes, mannequins, tires, oil drums, window frames, bathtubs, freezers, water heaters, ashcans and a mobile home trailer.

At its peak in the early 1970s, the site received over 40,000 visitors during the open months of April and December. As Greco's health declined and mass de-industrialization ravaged the city, Holy Land, U.S.A. followed suit. It did not re-open for the April 1984 season and the deterioration that had begun continued unabated for three decades. It was purchased in 2013 and a renewal effort is underway, with hopes of restoring it as a devotional site. While its future is up for grabs, the circulation of Holy Land, U.S.A. as a sign in public discourse has been dominated by two interpretive frames: kitsch and folk art.

In the *kitsch* frame, it is gawked at, a source of bemusement or ridicule. It is a spectacle, an oddity, fascinating for its utter strangeness. It appears regularly in books like 2005's *Weird New England*. A popular travel guide, *Roadside America*, once described Holy Land, U.S.A. as a 'post-nuclear, *Road Warrior* vision of the Holy Land. Most of the rambling spread consists of impenetrable assemblages of junk' (Barth et al. 1986: 154). In November 2002, the site was spoofed by Stephen Colbert on *The Daily Show*. In his characteristic style, Colbert dons a tan cargo vest to explore 'a religious Epcot'. In this *kitsch* frame, sites are fun to engage with, but worthy of lampooning not aspiration.

An alternative frame for understanding Holy Land, U.S.A. is to attribute it the status of *folk art*. In this frame, the site has value because it represents the creative work of a 'folk', visionary' or 'outsider' artist. The preferred term varies, but the spirit remains: someone who expresses their artistic vision through unpredictable media and without formal training. The subject matter is far less important than the location, the style and the relation between the artist's work and their biography (cf. Promey 2018). *The Clarion*, published by the American Folk Art Museum, featured a profile of Holy Land, U.S.A. soon after its peak popularity. While critical of the theology, the author praised the artistry: 'I felt that I was sharing an inner-vision made real, a manifestation of a man's lifelong dream come true and it was this feeling that made me exclaim that surely this was the real thing' (Ludwig 1979: 31).

By tracing the circulation of sites as signs we come to better understand the dynamics of contested public meanings. This calls attention to the fact that sites themselves are not the only actors that make claims of naming and defining, and that the social life of sites is crafted by many hands.

Coda

The Bible and Global Tourism seeks to be a rich and generative resource for understanding the diverse expressions of what this volume terms biblical tourism. Here, I have offered a further framework for reading across the chapters and conceptualizing what is to be gained from closely analyzing biblical tourism. The proposed framework centers on processes of circulation – the embodied, material and discursive ways in which biblical tourist sites have social lives beyond local emplacement and work together as an uncoordinated network.

This volume complements other efforts in the comparative study of pilgrimage and tourism to de-center individual traveler motivations and elevate the infrastructures, social exchanges and cultural conditions that direct and orient travel experiences (e.g., Coleman and Eade 2004). Self-identifications and brandings as pilgrim/pilgrimage, tourist/tourism or, remembering our opening example, pioneer/expedition are certainly fascinating, but taken alone limit the inquiry to questions of desired and imposed identities. We must contextualize the social play of how people understand themselves and how others seek to classify them amid broader processes of historical accumulation, transformation, erasure and authorization. Circulation, like mobility and contestation, fosters this more omnivorous inquiry.

Notes

1. https://www1.cbn.com/cbnnews/israel/2019/october/saudi-arabia-to-allow-access-to-ancient-biblical-sites-including-lsquo-the-real-rsquo-mount-sinai.
2. For example, see Richardson 2019.
3. https://livingpassages.com/.
4. https://livingpassages.com/real-mount-sinai-forgotten/.
5. The following is based on an interview with Bill Warren conducted by the author in November 2017, and a reading of the filed legal brief: 'Warren et Ux v. Collier Twp. Bd. Of Comrs' (1981).

References

Badone, E. (2007), 'Echoes from Kerizinen: Pilgrimage, Narrative, and the Construction of Sacred History at a Marian Shrine in Northwestern France', *Journal of the Royal Anthropological Institute*, 13: 453–70.

Bartal R., N. Bodner and B. Kuhnel, eds (2017), *Natural Materials of the Holy Land and the Visual Translation of Place, 500–1500*, London: Routledge.

Barth, J., et al. (1986), *Roadside America*, New York: Simon & Schuster.

Barush, K. R. (2018), 'Enacting the Glastonbury Pilgrimage through Communitas and Aural/Visual Culture', *International Journal of Religious Tourism and Pilgrimage*, 6 (2): 24–37.

Bielo, J. S. (2017), 'Performing the Bible', in P. Gutjahr (ed.), *The Oxford Handbook of the Bible in America*, 484–503, New York: Oxford University Press.

Bielo, J. S. (2018), 'Flower, Soil, Water, Stone: Biblical Landscapes Items and Protestant Materiality', *Journal of Material Culture*, 23 (3): 368–87.

Bielo, J. S. (2020a), 'Experiential Design and Religious Publicity at D.C.'s Museum of the Bible', *The Senses and Society*, 15 (1): 98–113.

Bielo, J. S. (2020b), '"Where Prayers Where May be Whispered": Promises of Presence in Protestant Place-Making', *Ethnos*, 85 (4): 730–48

Citro, Joseph A. (2005), *Weird New England: Your Travel Guide to New England's Local Legends and Best Kept Secrets*. New York: Sterling.

Coleman, S. (2014), 'Pilgrimage as Trope for an Anthropology of Christianity', *Current Anthropology*, 55 (s10): s281–s91.

Coleman, S. and J. Eade, eds (2004), *Reframing Pilgrimage: Cultures in Motion*, London: Routledge.

Engelke, M. (2013), *God's Agents: Biblical Publicity in Contemporary England*, Berkeley: University of California Press.

Eyl, J. (2019), 'Anachronism as a Constituent Feature of Mythmaking at the BibleWalk Museum', in E. Roberts and J. Eyl (eds), *Christian Tourist Attractions, Mythmaking, and Identity Formation*, 111–27, London: Bloomsbury.

Handman, C. (2018), 'The Language of Evangelism: Christian Cultures of Circulation Beyond the Missionary Prologue', *Annual Review of Anthropology*, 47: 149–65.

Jackson, J. (2013), *Thin Description: Ethnography and the African Hebrew Israelites of Jerusalem*, Cambridge, MA: Harvard University Press.

Kaell, H. (2012), 'Of Gifts and Grandchildren: American Holy Land Souvenirs', *Journal of Material Culture*, 17 (2): 133–51.

Kaell, H. (2014), *Walking Where Jesus Walked: American Christians and Holy Land Pilgrimage*, New York: New York University Press.

Long, B. O. (2003), *Imagining the Holy Land: Maps, Models, and Fantasy Travels*, Bloomington: Indiana University Press.

Ludwig, A. I. (1979), 'Holy Land, U.S.A.: A Consideration of Naïve and Visionary Art', *The Clarion*, 28–39.

Morgan, D. (2014), 'The Ecology of Images: Seeing and the Study of Religion', *Religion and Society: Advances in Research*, 5: 83–105.

Promey, S. (2018), 'Testimonial Aesthetics and Public Display', *The Immanent Frame*, 8 February, https://tif.ssrc.org/2018/02/08/testimonial-aesthetics-and-public-display/.

Rogers, S. S. (2011), *Inventing the Holy Land: American Protestant Pilgrimage to Palestine, 1865–1941*, Lanham, MD: Lexington University Press.

Richardson, J. (2019), *Mount Sinai in Arabia*, WinePress Media.

Stephenson, B. (2010), *Performing the Reformation: Public Ritual in the City of Luther*, Oxford: Oxford University Press.

Wharton, A. J. (2006), *Selling Jerusalem: Relics, Replicas, Theme Parks*, Chicago: University of Chicago Press.

Wimbush, V., ed. (2015), *Scripturalizing the Human: The Written as the Political*, London: Routledge.

Index

African Hebrew Israelites 262
Alps 34, 41
Altdorfer, Albrecht 19
American Bible Society 192
American Folk Art Museum 268
Anglican (church) 33, 171
Archaeology 8, 40, 68, 71, 160, 163–4, 179, 183–6, 188–9, 195, 197, 262
Architecture 6, 57, 150–1, 209, 245, 250–3
Authenticity 3, 10, 141, 167–8, 217, 250, 254, 256, 262
Authority and Authorization 3, 7, 8, 9, 31, 34, 64–5, 76, 85, 92, 118–19, 184, 203, 217, 263–4, 266–7, 269

Baptist 65–6, 69, 71–2, 75, 90, 181
Biblel + Orient Museum (Freiburg) 196
Bibelhaus Erlebnis (Frankfurt) 185
Bibelwelt (Salzburg) 186
Bible Lands Museum 186, 193
Bible Museum (Goodyear) 182
Bible Walk (Ohio) 265–6
Bibleworld Museum and Discovery Centre (Rotorua) 183
Bijbels Museum (Amsterdam) 188
British and Foreign Bible Society 182
British Isles 24–5, 28–9
British Library 191
British Museum 215

Callaway, Elvy E. 63, 65–6, 73–8.
Camino: de Santiago 2, 10, 13, 124–144, 149–50, 266; Francés 124–5, 130, 142, 150
Cantwell Smith, Wilfred 118
Catholicism and Catholics 2, 10, 12, 33, 47, 50, 125, 127, 133, 161, 195, 209–10, 217, 222, 226–7, 262–3, 265, 267
Chabad (see Judaism)

Christian Zionism 188
Christiansfeld (Denmark) 6, 12–3, 242, 244, 247, 249, 256, 263, 265
Church of the Holy Sepulcher 48, 51
Circulation 19, 263–9
Colbert, Stephen 268
Coleman, Simon 119, 123, 126, 140
Communitas 3, 119, 132–3, 135
Constantin de Groot 210–11
Cook, Thomas 40–1, 43, 45–8
Creationism: Answers in Genesis 201, 215; Ark Encounter 190, 203, 214, 217; and biblical literalism 201, 210; Creation Museum (Kentucky) 184, 201, 215, 267; and natural history museums 11, 202; safaris 261; and science 8, 204; visual culture of 202, 215–8

Darwin, Charles 11, 64, 67, 201
Dead Sea Scrolls 186–7
Dutch East India Company 212

Eastern Orthodoxy 147, 160, 168, 186, 261, 263
Egypt 39–41, 66, 92, 216, 261–2, 265
Eliade, Mircea 5, 119, 231
Embodiment 6, 74, 93, 119, 141, 222, 225, 269
Erasmus, Desiderius 17, 28
Estevez, Emilio 127
European Cultural Routes 8, 10, 147–9, 150–5, 166, 265
Evangelicalism 64, 162, 180, 187–8, 193, 195, 203, 261–2, 267
Evolution, theory of 11, 73, 201–2, 214

Fairfield Warren, William 63, 65–9, 71–3, 76–7,
Festival 5, 7, 11–3, 109, 221–30, 233–9, 254, 263, 264

Frankfurt Bible Museum 188
Friends of Zion Museum (Jerusalem) 188

Garden of Eden 2, 5, 9, 62–6, 70, 73–5, 77–8, 91, 184–5,
Garden of Hope (Kentucky) 267
Globalization 4, 109, 224, 238
Golden Legend (Legend Aurea) 20, 123
Gould, Stephen Jay 202, 208
Grand Tour 42
Great Passion Play (Eureka Springs) 181
Guidebooks 7, 9, 13, 38–45, 48, 54–7, 124, 129, 143, 265–7

Healing: 106, 108, 136–7, 140, 144, 263–4; and biomedicine 115, 118; stories of 8, 110, 112–18, 266
Hebrew Bible 42, 45, 49, 52, 55, 57–8, 70, 187–9, 265
Heritage: 1–3, 5–12, 86, 125, 147–8, 159, 163, 221–2, 236, 239, 242, 246, 249–50, 252–3, 256–7; process of heritagization 9, 11, 263, 266; religious 84, 91, 93, 95–7
Holy House of Nazareth 6, 9, 21–5, 27–9, 31–4, 262
Holy Land 2, 5, 9–10, 17–8, 20–2, 29, 31–4, 39–41, 44–51, 54–5, 57–8, 83–8, 97, 105, 182, 186, 188, 197, 209
Holy Land, U.S.A. (Waterbury) 267
Holy Land Experience (Orlando) 180, 196
Holy longing 4, 5

Inter-religious dialogue 150, 228
Islam 20, 41, 45, 58, 62, 105, 187–8, 228, 261

Jaffa 40–1, 46, 56–7
James, St. 24, 123–4, 140, 149, 151, 266
Jebel Musa 261
Jerusalem 17, 20–2, 40–1, 45–53, 55–7, 86–7, 93, 151, 156, 158, 183, 185–8, 193, 265
Jewish Museum (Berlin) 194
Judaism: 10, 49–50, 61, 104–7, 228; Hasidism 105–7, 262; Jewish(ness) 2, 5, 9, 12, 47–58, 104–5, 107, 109–10, 112, 115, 155, 186–8, 194, 210–1, 261; Lubavitch 106; and scripture 190

Kemp, Wilfred 134–8
Kerkeling, Hape 130–3, 138, 142

Latour, Bruno 204, 207–9, 211, 217–8
Latter-Day Saints (LDS) (see Mormonism)
Leisure 3, 39, 44, 125, 144, 216
Liminality 128, 135
Lived religion 2, 242–4, 246, 249, 252–3, 256–7, 259
Living Torah Museum 186
London (United Kingdom) 27, 46, 86, 112, 181, 191–2
Loreto (Italy) 6, 8, 21–2, 24–6, 29–34, 262, 266
Lyell, Charles 201

Maastricht 11, 225–9
Maps 7, 40, 65, 140, 157, 164, 181, 186, 207, 211, 214, 249, 250
Material culture: 9, 242, 253–4, 265; pewter badges 13, 18–19, 21, 24, 28, 31, 33–4; relics and reliquaries 18, 21–2, 26, 29, 31–2, 34, 192; souvenirs as 9, 33, 48, 89, 265
Mecca 39
Media 6–7, 10, 33, 106, 124, 126, 133, 137, 143, 144, 164, 194, 219, 227, 261, 264, 266–8; Mass media 2, 9; and pilgrimage 124, 126–7, 144; and tourism 126–7
Memory and Memorialization 2, 6, 8, 83, 93, 150, 159, 169, 266
Methodist 63, 65–8, 193
Michelangelo 204–6, 208–9, 215
Migration 4, 52, 87, 92–3
Miracle 19, 22, 27, 104–8, 110–1, 113, 115–7, 119, 124, 204, 213–4, 218, 230–1
Moravian Church: 5, 6, 12, 242, 244, 248, 251–2; rituals of 254–7
Morgan, David 7, 202, 209–10
Mormonism: 64, 83–5, 88, 93, 96–7, 262; Book of Mormon 10, 85, 89–90, 96; Mormon Trail 92–3
Museum of Biblical Art 192
Museum of the Bible (Washington, D.C.) 181, 184, 188–9, 192, 194–6, 263
Museum of the Book (London) 181, 192

Museums: 1, 5–8, 11, 13, 49, 70, 74, 76, 151, 155, 165, 221, 252, 256–7; definition of 180; and religion 243; and secularity 189; typology of 180–93
Music 96, 162, 222–39
Musica Sacra Maastricht 7, 222–4, 226, 228–9, 232, 235–6

Nationalism 186, 188
Nauvoo Temple 92
Nazareth (Israel) 7, 18, 20–3, 26, 30, 32, 184, 263
Netherlands, the: 1, 6, 11, 24, 127, 188, 194, 212, 221; Low Countries, the 29, 32
New Jerusalem 86, 91–2
Nooteboom, Cees 143
North Pole 10, 63, 66–9, 72

Ohel (Lubavitcher Rebbe burial site) 104, 110, 266
Orientalism 186, 262
Otto, Rudolf 227, 231, 235

Palestine 2, 21, 39–48, 51–2, 54–6, 86, 156, 188, 262, 265
Palestine Park 265
Palmyra (New York) 89–90
Paradise 62–3, 65–6, 69, 72–4
Pärt, Arvo 228, 234
Paul (biblical apostle) 8, 11, 92, 147, 155, 186, 261, 265
Pennsylvania German Cultural Heritage Center 189
Pilgrimage: and healing (miracles) 104, 106, 110, 136–7; and Judaism 104–5, 118; and relation to tourism 2–4, 6–10, 17, 20–2, 29, 32–4, 39–40, 42, 47, 57–8, 78, 83, 87, 95, 97, 104, 161, 244, 269; and ritual process 91, 111, 132; as theme in the anthropology of religion 104
Photograph(y) 39–40, 45, 49, 53, 131
Postsecular 7, 11–2, 221–3, 239, 264
Protestant(ism) 2, 5, 11–2, 47, 63, 65, 161, 190, 195, 204, 218, 237, 250, 261–3; visual culture of 209–10, 213–4
Pyrenees 124, 128, 138,

Reformation 32–3, 181, 190, 263
Rome 21–2, 48, 50, 53, 152, 156, 158, 193, 262

Sacred, the: 3–4, 7, 11–2, 89, 106, 111, 119, 130, 218, 222–3, 226–9, 231–6, 238–9, 244–5; Sacrality 7, 242; Sacralizing 83, 96, 266
Saint-Jean-Pied-de-Port 130, 138, 142–3
Salt Lake City (Utah) 96
Santiago de Compostela (see Camino de Santiago)
Schneerson, Menachem Mendel 10, 104, 107, 109, 114, 262
Science 8, 11, 38, 68–9, 76–7, 201–2, 208–9, 218, 262
Scripture 2, 5, 6, 9, 11–3, 17, 45, 64–5, 70, 84–5, 147, 183, 203, 211, 213, 215, 217–8, 223, 243, 251, 262; the social life of 2, 10, 13, 105, 118–19, 262, 266–7
Secular, the 133, 193, 222, 223, 252, 257, 266; Secularization 4, 12, 127
Shrine of the Book (Jerusalem) 186
Silence 234, 237
Sistine Chapel 204–6, 208
Smith, Joseph 64, 84–6, 89–92, 97

Technology 39, 105, 118, 227; Digital technology 4, 180, 264
Temple Mount 49, 57, 188
Temple of Solomon (Brazil) 265
Temporal(ity) 5–7, 39, 225, 266
Tersatto (Croatia) 21
Thorvaldsen, Bertel 93–4
Tourism: industry of 28, 147, 164–5, 168, 223, 254, 261; and religion 2, 9–10, 12, 97, 104, 165, 168–9, 243, 245; slow 1
Tree of Knowledge 63, 73
Tree of Life 62, 68
Turner, Victor and/or Edith 3, 119, 128, 132

UNESCO 1, 5, 8, 12, 125, 147, 160, 162, 166, 242–6, 249, 253, 257, 261
Uniformitarianism 201–2, 214

Vatican Modern Sacred Art Museum 217
Via Dolorosa 50
Voragine, James of 20, 123–4

Walsingham (England) 6, 18, 25–9, 31–4, 266
Wavre (Belgium) 6, 18, 29–32, 34, 266

West, Rev. Edmond Landon 63, 65, 69–73, 76–7
Western Wall 49, 52–3, 57

Yehoash 52–4
Young, Brigham 87, 91–2

www.ingramcontent.com/pod-product-compliance
Lightning Source LLC
Chambersburg PA
CBHW072129290426
44111CB00012B/1831